THE GREAT DINOSAUR EXTINCTION CONTROVERSY

THE GREAT DINOSAUR EXTINCTION CONTROVERSY

Charles Officer & Jake Page

HELIX BOOKS

ADDISON-WESLEY PUBLISHING COMPANY, INC.
Reading, Massachusetts Menlo Park, California New York
Don Mills, Ontario Harlow, England Amsterdam Bonn
Sydney Singapore Tokyo Madrid San Juan
Paris Seoul Milan Mexico City Taipei

Library of Congress Cataloging-in-Publication Data
Officer, Charles B.
  The great dinosaur extinction controversy / Charles Officer and Jake Page.
    p.   cm. — (Helix books)
  Includes bibliographical references and index.
  ISBN 0-201-48384-X
  1. Extinction (Biology)   2. Dinosaurs.   3. Cretaceous-Tertiary boundary.   I. Page, Jake.   II. Title.
  QE721.2.E97036   1996
  567.9'1—dc20   96–3809
                                                                        CIP

Jacket design by Jean Seal
Text design by Karen Savary
Set in 11-point Calisto
by Shepard Poorman Communications Corporation

1 2 3 4 5 6 7 8 9 10-MA-0099989796
First printing, June 1996

*For every complex problem there is a solution
which is simple, neat, and wrong.*

—H. L. MENCKEN

*To connect the dinosaurs, creatures of interest to
but the veriest dullard, with a spectacular event
like a deluge of meteors . . . seems a little like one
of those plots that a clever publisher might concoct
to guarantee enormous sales. All [the theories]
lack is some sex and the involvement of the Royal
Family and the whole world would be paying
attention to them.*

—IAN WARDEN

CONTENTS

PREFACE

IN THE AUTUMN OF 1994, THE INEVITABLE OCCURRED.
It had occurred to some television producers that if Orson
Welles could pull off a Martian invasion of New Jersey in the late
1930s, they could pull off the destruction of Planet Earth by a hail
of incoming meteorites just a few years before the turn of the
calendar into the next millennium. Of course, the Welles invasion
had caused an actual panic among a gullible public that tuned into
his radio program, and the TV producers (probably with advice
from counsel) had no desire to do *that*. So when the docudrama
aired, about a week before the 1994 midterm elections, it was
punctuated by announcements that what the viewer was seeing on
TV was not *really* happening. Even so, there were subsequent
reports that large numbers of people did believe it, though they
evidently suffered their panic in the privacy of their homes; there
was no Wellesian taking to the street.

For the sake of dramatic verisimilitude, the TV producers
hauled newsman Sandy Vanocur out of retirement and installed
him in the New York headquarters of *World Network News*, where
he was joined by an exhausted-looking woman science editor as

the news began to emerge that Earth had been struck by not one but three large meteorites.

Virtually all sentient Americans in 1994 had heard repeatedly that a giant meteorite wiped out the dinosaurs some 65 million years ago, and many believed the Earth was imminently threatened by another such catastrophe. The National Aeronautics and Space Administration (NASA) and various astronomers had been warning about this threat for a number of years. The case for "sentinel" telescope arrays, indeed, seemed plausible to many reasonable people.

So prevalent in the public mind is the notion of the meteoritic "big one," waiting out there to strike (just as the seismic "big one" is part of the subconscious of Californians), that the TV producers must have felt that the watching public was already blasé. So, they invoked not one but three meteorites, launched (however improbably) as a harmless, attention-getting overture by extraterrestrials. This was followed by President Clinton's misguided decision to shoot the next ones down with nuclear missiles launched from fighter-bombers streaking toward the north pole. Predictably, the ETs were incensed and launched a massive meteoritic counterstrike. Sandy Vanocur and his colleague at the news desk said a final and brave farewell to human civilization (which presumably had nothing better to do during Armageddon than keep watching the television set).

Now this is a book about science—about meteorites, dinosaurs, and a host of other matters—and not about popular culture, good or bad. But dinosaurs *are* a part of popular culture and have been to one degree or another since their discovery in the nineteenth century. The film *Jurassic Park* is only the latest manifestation, along with the ubiquitous robodinosaurs that grace museums and theme parks alike. And meteorites have been a (perhaps less marketable) part of popular culture ever since science fingered at least one as the culprit in the great dinosaur massacre, or more properly, extinction.

It was in 1980—only a decade and a half before we set out to write this book—that Nobel Prize–winning physicist Luis Alvarez announced his theory that the cause of the dinosaurs' demise was extraterrestrial—a gigantic meteorite. Within a few years his theory was more than generally accepted. It was common knowledge—*a given*. (As late as 1994, a friend of junior author Page and editor of *The Skeptical Inquirer*, a journal devoted to puncturing pseudoscientific theories and assertions, said that Page was "the only person on Earth" who didn't believe that a meteorite impact wiped out the dinosaurs.) In fact, the impact theory is startling and radical and became accepted faster than any other that we can recall. It took a half-century, for example, for the notion of continental drift to become accepted by a majority of geologists, much less anyone else, and as late as the 1980s an only partly tongue-in-cheek Stop Continental Drift Society still flourished.

In any event, suddenly dinosaurs were big-time again in the halls of science itself. Dinosaurs had for a long time taken a back seat to other less dramatic but more rewarding fossil studies. Perhaps in part because they were so fascinating to children, dinosaurs were generally considered "kid stuff," even among many paleontologists. Certainly they were pretty trivial in the minds of high-energy physicists, molecular biologists, and researchers in other heroic areas of science that, along with medical research and space, garner the lion's share of federal science dollars.

It is tempting to point out that for several years now we have been approaching a millennium. The years approaching the previous one, the year 1000, were evidently a time—in Europe, at least—of widespread malaise, with many apocalyptic prophecies taking flight and large-scale lunacy abroad in the land. Those were more superstitious times, of course, than ours, but in our own time we too appear to be facing a widespread malaise. There is a generalized fear in many minds that, in spite of unprecedented technical progress, the wheels of civilization are falling off. Bizarre apocalyptic cults are heard from, to be sure, but this malaise is nowhere better observed than in the concern over the environment. If such self-inflicted problems are not enough to make us uneasy, mankind now has the ghosts of the dinosaurs as a stern warning of the havoc that can be wrought on even an innocent world by an extraterrestrial (albeit perfectly natural) cause.

In the public mind The Meteorite lurks out there now like a buzzing in the ears—if not an agent of cosmic vengeance for humanity's hubris, then at least a catastrophic accident waiting to happen. In 1994 a University of New Mexico astronomer got on the radio to advise people where and what to look for in the night sky at the time of the annual Perseid meteor shower. He added that it was perfectly safe and no one "probably" would be struck by an incoming meteorite. This is a form of alarmism, however negatively stated, and it seemed irresponsible coming from a scientist's lips.

In 1980, at the birth of the Alvarez impact hypothesis, a majority of scientists were profoundly skeptical of it. The voices of these skeptics (most of them from paleontologists and geologists who had been working in these fields for years) tended not to be heard, and they complained, often bitterly, among themselves and to any science editor or journalist whose ear they could catch. They went largely unheeded outside the confines of their own disciplines and the technical journals devoted to those disciplines.

Senior author Officer, a Dartmouth geologist, is one of these original skeptics. Page is a science writer and former museum science editor. The two of us collaborated earlier on a book, *Tales of the Earth*, that detailed just how much havoc the Earth itself can cause in the course of its own metabolic rumblings. We tried to put anthropogenic havoc (what humans are doing to the planet) in that context. We have always inclined to the notion that Mother Earth—the impresario of a long-running evolutionary play in an ever-changing ecological theater—has plenty of her own techniques for colluding in the extinction of whole species, even dinosaurs.

In this book we look at the meteorite impact theory as it "evolved" over its short life, and at the scientific efforts, both good and bad, that followed its announcement. And we give an alternative explanation, one that existed before 1980 and has been further enriched in the course of the fracas. We conclude that the Earth is the major heroine here, a heroine that may not be as dramatically sexy as the meteorite theory but one that is also without need for that rather weak and desperate ploy of playwrights, a deus ex machina.

In a sense, this is a shame. Both of us like a bit of drama with our breakfast cereal as much as the next person. But the impact theory has not stood up at all to expert scrutiny. Actually, many theories do not. In the history of science, theories and hypotheses are just that—interesting possible explanations of observed facts that can be tested, either in their own right or by weighing them against others. Among the many hypotheses that prove false, some have eventually been useful in furthering scientific understanding. But the impact theory has not been helpful even in this regard.

Indeed, most of the "science" performed by the Alvarez camp has been so inexplicably weak, and the response to it so eagerly accepting by important segments of the scientific press (never mind the popular press and the tabloids), that some skeptics have wondered if the entire affair was not, on the impact side, some kind of scam. But such an allegation would have to be backed up by the kind of investigative reporting that exposed the Watergate scandal, and we are not qualified to do that. We have instead confined ourselves largely to addressing the scientific merits of the case. We think this area tells a compelling enough story in itself.

The demise of the dinosaurs (along with a host of other more or less contemporaneous life-forms) clearly occurred for other and more complicated reasons than impacts from space. These reasons are wholly plausible and well understood, if not down to the last detail; t's and i's remain to be crossed and dotted. While this is increasingly apparent within the scientific community, it will take a longer time for the semipopular and popular press and therefore the general public to understand and accept this. The extinction of entrenched lore after all—even lore that has arisen with such meteoric rapidity—is usually a drawn-out affair. This book, we hope, is an accessible and helpful shove in the right direction.

ONE
The Day of the Meteorite

THE CITIZENS OF GUBBIO, ITALY, ABOUT A HUNDRED miles due north of Rome, may take some pride that their town's name is associated with one of the most electrifying scientific hypotheses in Earth history. It is not every decade that a geological hypothesis arises that commands such widespread popular attention as to make, say, the cover of *Time* magazine. But it was near Gubbio in the late 1970s that a geologist named Walter Alvarez of the University of California at Berkeley came across an unusual layer of clay.

It was a thin layer, only a few inches thick, and on analysis it was found to contain a tiny amount—in fact, a mere 9.4 parts per billion—of a rare element known as iridium.

Iridium, element number 77 on the Periodic Table of Elements, is one of the so-called platinum group, all of which are relatively rare. Besides platinum and iridium, the group includes osmium, palladium, rhodium, and ruthenium—a clan of inert, metallic elements. Iridium is white, extremely hard, and resistant to acids, and it has been found useful in hardening platinum yet further for use in pen points, scientific apparatuses, and surgical tools. But it is rarely used because it is rarely found. It has not been found at all in that great global solvent, seawater, for example, and in the Earth's crust it normally occurs at less than 0.03 parts per billion. So in the Gubbio clay beds, Alvarez came across a thin layer that contained *three hundred times* the expectable amount of iridium!

That thin iridium-rich layer occurred stratigraphically at or near the boundary between two great geological periods—the Cretaceous and the Tertiary, some 65 million years ago—widely known to be the time when a great host of life-forms disappeared from the planet, mostly vast populations of teeming microscopic marine organisms, but also, and most notably, the dinosaurs.

In any event, Walter Alvarez's find in Gubbio set the stage for an important scientific announcement with his father, Luis, as

Luis Alvarez. *Source:* Lawrence Berkeley Laboratory, University of California.

senior author and impresario. To be brief, Alvarez *père* was a Nobel physicist of great achievement in several realms, including high-energy physics, and was widely respected for having the clear mind of an inventor. He, his son, and two other investigators realized that a ready source for such an abundance of iridium could be found in outer space. Meteorites, those occasional stony or metallic visitors that scream in from space, lighting up the sky with a fiery trail—most of them former denizens of the asteroid belt between Jupiter and Mars—were known generally to contain 500 parts per billion of iridium.

And so, on June 6, 1980, *Science* magazine, the weekly publication of the American Association for the Advancement of Science, published an article by Luis Alvarez, Walter Alvarez, Frank Asaro, and Helen Michel entitled, "Extraterrestrial Cause for the Cretaceous-Tertiary Extinction."

The article was couched in scientific language, which is typically as dry as dust from the Moon, and most of its thirteen pages described various geophysical and chemical matters that—to laypeople and to many scientists from other disciplines—were almost as dry. But embedded in the abstract to the paper, in proper passive voice, were the following words:

> A hypothesis is suggested which accounts for the extinctions and the iridium observations. Impact of a large earth-crossing asteroid would inject about 60 times the object's mass into the atmosphere as pulverized rock; a fraction of this dust would stay in the stratosphere for several years and be distributed worldwide. The resulting darkness would suppress photosynthesis, and the expected biological consequences match quite closely the extinctions observed in the paleontological record.

Here was a bombshell. In fact, the Alvarez group suggested that a meteorite ten kilometers (about six miles) in diameter, weighing 10^{10} tons, struck the Earth at thousands of miles per hour. It would have blown out a 200-kilometer crater—a dent in the Earth's crust 120 miles across (a two-hour drive on a good highway)—and sent sixty times its mass into the atmosphere to create a rapidly

Earth-enveloping cloud of fine particles, including an inordinate amount of iridium derived from the meteorite's body.

Before long, the Alvarez group, with Alvarez *père* still in the role of senior investigator, suggested that infrared radiation (heat) could have penetrated this dust layer. So that during the years when the dust layer hid the sun, heat would nonetheless have escaped, chilling the Earth to the approximate conditions of a wintry night—that is, cold *and* dark.

To return to June 6, 1980: *Science* is not read by many lay-people, but it has a wide and devoted readership in the scientific community. For many scientists, the Alvarez hypothesis came as a thing of beauty: First, it contained some new and unexpected data. Second, it provided an exciting and, importantly, neat idea for explaining some heretofore mushy matters. And third, it was testable.

The hypothesis was also extremely exciting, of course, to the public. Also among *Science*'s readers are virtually all who would call themselves serious science journalists, and they leaped on this study with unbounded enthusiasm. Nowhere would the media appeal of this hypothesis be better summed up than, a bit later, by an Australian journalist, Ian Warden, in *The Canberra Times*:

> To connect the dinosaurs, creatures of interest to but the veriest dullard, with a spectacular event like the deluge of meteors . . . seems a little like one of those clever plots a publisher might concoct to guarantee enormous sales. All [the theories] lack is some sex and the involvement of the Royal Family and the whole world would be paying attention to them.

Indeed, practically the whole world did. Not only did the theory make the cover of *Time*, but wonderfully lurid illustrations of doomed dinosaurs staring dumbly into a sky lit by a vast fireball appeared on the cover of virtually every supermarket tabloid and kids' comic book and in every magazine rack. Popular magazines devoted to science—such as *Omni* and *Discover*—took it up. Within a mere historical instant, the notion was in the lay mind a fait accompli, an unquestioned given, and not merely a hypothesis awaiting that normal fate of all scientific hypotheses—that is to say, skeptical and reasoned criticism from other scientists, by which it could be proven, refined, or dismissed. Even within the realm of scientific debate, the Alvarez hypothesis did not follow the normal procedures of science. Dissent was immediate—but it went largely unreported, even in much of the scientific press. Dissenters were in fact publicly excoriated, even ridiculed, by the hypothesis's proponents. Rancor seethed in scientific halls and meetings. This was not the gentlemanly and collegial debate that usually lies at the heart of scientific discourse. In the process, a great deal of very bad scientific thought and procedure occurred—not just bad behavior. Alvarez proponents simply ignored a vast amount of data that other scientists had patiently collected over the decades and continued to collect—evidence that increasingly clearly gainsaid the new hypothesis.

The Alvarez hypothesis and the gauntlet it threw down also contributed to generating a great deal of good science. And it did focus, however waywardly for a time, scientific and public attention on an issue that even the "veriest dullard" among us had begun to sense was a matter of increasing urgency: mass extinctions.

Extinctions of species are the sometimes-unappreciated background buzz of life and evolution. More species have winked out forever than now exist on the planet. Most extinctions were localized if fairly common events that have occurred throughout the history of life—paleontologically no big deal. Extinct species are typically survived by related forms. Today we have no more mastodons, but at least for now, we have elephants. By contrast, mass extinctions tend to leave little or no trace of the creatures that existed before; these are catastrophic shifts as well as poignant matters of life and death. In the case of the dinosaurs, after some point that was largely coincident with the Cretaceous-Tertiary (or what is called K-T) boundary, there simply were no more of them.

Given much of the environmental news in the latter decades of this century, many people are aware that we may be facing a period of phenomenal extinction of species. The rates of extinction among mammals and birds may seem relatively low, but a little arithmetic shows how perilous life is these days. Of the approximately 4,200 existing species of mammals, an estimated sixty-three have become extinct since A.D. 1600—all, so far as is known, at the hands of humans. In the same interval, eighty-eight of some 8,500 species of birds have gone extinct. This is one and a half percent of all mammal species and one percent of birds—low percentages until they are carried forward. If such rates—say, one percent every four hundred years—persist, *one hundred percent* of all mammal and bird species will be forever gone in a mere 40,000 years—a very short interval in geologic time. And the denouement may not take even that long, for rates of extinction appear to be increasing as habitat destruction proceeds apace around the world, taking with it uncounted species that do not have as loyal a following as, say, pandas and dusky seaside sparrows. To many, it is clear that we are currently inside the gate of a vast and deep extinction crisis. That this highly complex and planetwide affair is almost entirely the result of human activity is now widely understood and accepted. One of the unfortunate aspects of the publicity attending the Alvarez hypothesis is the extent to which its extraterrestrial dramatics may have diverted attention away from the anthropogenic havoc now under way in the biological world.

It is hard enough to understand the present; it goes without saying that to understand events that took place in the far distant past—for example, millions of years ago—is all the harder. Such events, particularly extinction events, are part of a grand mystery story, a tale littered with corpses and potential perpetrators. Of all extinctions in the past, that of the dinosaurs has commanded the most attention in the popular mind, and in certain circles and at certain times, it has commanded a great deal of scientific attention. The issues raised by the Alvarez hypothesis go far, far beyond the matter of the dinosaurs and range widely into many other biological realms and many other branches of science. These will be explored in the following chapters, but dinosaur extinction theories have a history of their own.

The first dinosaur remains were uncovered and described about two hundred years ago. Initially, they were considered those of some kind of unusual lizard. At the 1841 annual meeting of the

British Association for the Advancement of Science, anatomist and paleontologist Richard Owen first introduced the term *Dinosauria,* meaning "terrible lizard"—to describe them. Owen argued that God had chosen the Mesozoic era, the geologic era in which dinosaurs lived, as suitable for them because of its low atmospheric oxygen level. As reptiles, he said, they had low metabolic rates and could survive on a lower oxygen intake than, say, present-day mammals. Oxygen levels had risen during the Mesozoic to a level that was unsuitable for dinosaurs, he argued, and as a consequence they died out. It was, of course, a circular argument; Owen had no evidence at the time about how (or if) oxygen levels had varied in the Mesozoic. But many of the causes of extinction proposed since his day have been, if anything, more outlandish.

Michael Benton, a paleontologist at the University of Bristol in England, provided an informative and at the same time amusing account of dinosaur extinction theories in a recent article in the journal *Evolutionary Biology.* (Not all scientific treatises are without the leaven of wit.) He divides the history of such theories into three phases. First, the "nonquestion" phase lasted until about 1920. It was succeeded by the "dilettante" phase, which extended up to about 1970 and was itself followed by the phase we are now in, the "professional" phase.

For much of the nineteenth century and early twentieth (the nonquestion phase), the prevailing view was that the dinosaurs simply petered out from racial senility. It was a widely accepted credo that certain groups of animals became moribund in their development, that their source of evolutionary energy and novelty dried up. If genetic variability is the gas that powers the ability to survive in a changing world, some creatures like the dinosaurs simply ran out of gas. The horns, spines, and other "frills" of late Cretaceous dinosaurs were cited as evidence of genetic senility—a kind of racial sclerosis or calcification.

During the 1920s dinosaurs and whatever might have done them in became a less serious topic among scientists. This was the start of the dilettante phase, when many nonpaleontologists came up with answers for what scientists took as an interesting but not very important question. It was proposed that dinosaurs had developed—à la Sigmund Freud, perhaps—a suicidal psychosis. One opiner said, "The Age of Reptiles ended because it had gone on long enough and it was all a mistake in the first place"—a twist on the racial senility theory. Somewhat less outlandish was the diet hypothesis. It was known that flowering plants—angiosperms—came into being just after the middle of the Mesozoic and gradually gained a larger and larger foothold, eventually to dominate the landscape. They presumably made up an increasing portion of dinosaurian diets but, being far less oily than ferns and conifers, caused plant-eating dinosaurs to die off from terminal constipation—thus, of course, depriving their carnivorous cousins of anything at all to eat.

The roster grew and grew. Mammals—small creatures—also arose in the Mesozoic, and as they became abundant, maybe they ate the dinosaur eggs. It was in the Mesozoic that insects began to proliferate—in part as pollinators of those angiosperms. Insects can carry plagues. After all, 50 percent of the population of Europe—and even more of the Asian population—was killed off by the bubonic plague, a flea-borne affliction. Couldn't some even more virulent plague have taken out the dinosaurs? Or maybe the climate grew too hot? Or too cold? Or too dry? Or too wet? There could have been a great rise in carbon dioxide levels. Or a great decrease. And so on.

After about 1970 paleontological circles decided that, such wild guesses aside, it would be good to know more about all this. Clearly there had been a massive die-off around the end of the Mesozoic at the K-T boundary. About 50 percent of all species had gone extinct—not just dinosaurs but much of the shallow-water shellfish and surface-dwelling plankton. Both kinds of marine organisms, it was realized, play an extremely important role in the overall economy (or perhaps better, ecology) of the oceans and therefore of life in general. And of course, 50 percent of species had survived. So these ancient extinctions were far more than a trivial matter.

As dinosaur extinction theories entered what Michael Benton called the professional phase, scientists looked for physical causes that could be corroborated by data in the geological record.

For example, there was evidence that the last of the dinosaurs had existed on the shores of a broad seaway that extended from to-day's Gulf of Mexico north to the Arctic Ocean, and that by the end of the Cretaceous, that seaway had dried up, later to be re-placed by today's Rocky Mountains. This suggested that the de-cline and demise of dinosaurs in at least western North America was the result of a *gradual* change in habitat and climate.

Another theory in this phase concerned a long period of im-mense volcanism that occurred over 400,000 years surrounding the K-T boundary. Evidence for such volcanism lies in the Deccan Traps, a vast area of flood basalt deposits in western India. The volcanic activity, it had been suggested, would put huge quantities of volatiles into the stratosphere—carbon dioxide leading to global warming, sulfur dioxide leading to acid rain, and chlorine leading to depletion of the ozone layer and increased ultraviolet radiation. In a geological sense, this disruption, however long it lasted in our terms, would qualify as a sudden *catastrophic* change.

It was into this phase of dinosaur extinction theory that the Alvarezes cast their hypothesis—postulating an almost instanta-neous catastrophe that purported to give a satisfyingly simple, im-mediate, and straightforward answer to the demise of the dinosaurs and other life-forms at this "moment" in geologic time. Indeed, Luis Alvarez himself could be seen as something of a deus ex machina, an experimental physicist—and one distinguished from virtually all his colleagues in the world by being a recipient of the Nobel Prize—bringing all the rigor of his field to a murky arena and, with wonderful flair, providing the needed light.

Robert Jastrow, for many years an astrophysicist with NASA and a professor at Dartmouth College in New Hampshire, once wrote in the magazine *Science Digest* about a pecking order that some believe characterizes the grand and far-flung enterprise of sci-ence. Jastrow recalled being told by a Nobel laureate in physics, then at Columbia, that "an intellectual hierarchy exists in science, with mathematics and theoretical physics on top, experimental physics just beneath, and then further down, chemistry and perhaps astronomy. Geology and paleontology, which deal with dirty ob-jects like rocks, are considerably lower on the list, and biology—at

least the parts that deal with soft, squishy things like entire organisms—is at the bottom. Anthropology, psychology, and the 'soft
sciences' like sociology are not on the physicist's list at all.''

Given this pecking order, people tend to take notice when a
physicist casts his rigorous mind on a more humble realm, and in
some cases such cross-fertilization has been enormously benefi-
cial. Without going into any detail, it is fair to say that molecular
biology, including genetics, would not have advanced so quickly
had not biologists been joined by physicists in the latter half of this
century. In a more political area, the Nobel chemist Linus Pauling
did the world a considerable favor by forcefully raising the issue of
radiation damage caused by atmospheric testing of nuclear weapons, which at least in part led to the 1963 ban on such tests and to
a Nobel Peace Prize for Pauling.

At the same time, cross-fertilizers can get into trouble—either
by being right (in the long run) but renounced at the outset, or by
being altogether wrong as a result of not understanding much of
the basics and methodologies of the field they are treading into
and making a jackass of themselves.

It was a meteorologist, Alfred Wegener, who brought to geology its greatest and most revolutionary concept in this century if
not of all time—continental drift or, as it is now called, plate tectonics. In his book, *Origin of Continents*, published in German in

Alfred Wegener. *Source:* **Historical
Picture Services, Inc., Chicago.**

1912 and in English three years later, he used current geological findings in support of his thesis that 200 million years ago there was no Atlantic Ocean, and North and South America were joined to Europe and Africa. The striking similarities in geologic structure, stratigraphy, paleontology, and petrology in the ancient rocks on either side of the Atlantic could best be explained if the two continental land masses had once been joined and had drifted apart, albeit at the imperceptibly slow rate of a fraction of an inch per year. The vast majority of geologists dismissed the notion as unscientific and inept. It contradicted the prevalent dogma that the continents and oceans have essentially always been where we find them: to suggest otherwise was to call for a wholesale rethinking of how the Earth works.

Nevertheless, in 1926 the American Association of Petroleum Geologists held a symposium on the topic. Wegener was invited but could not attend, and it was just as well, since the symposium was essentially an ambush. A professor from the University of Chicago said geologists might ask if theirs could still be called a science when it was "possible for such a theory as this to run wild." Another commented:

> It is not scientific but takes the familiar course of an initial idea, a selective search of the literature for corroborative evidence, ignoring most of the facts that are opposed to the idea, and ending in a state of autointoxication in which the subjective idea becomes objective fact.

Not just Wegener's theory but Wegener himself was attacked in absentia. It was by no means the first or last time that the dignified scientific arenas of objective investigation have rung with ad hominem attacks and even epithet.

The geological community and the rest of the world eventually came around to Wegener's view after 1963, when Fred Vine and Drum Matthews of Cambridge University in England published an article suggesting a plausible mechanism by which the drift of continents could have taken place. It was a matter of the seafloor spreading out from great underwater ridges like the Mid-Atlantic. There in the seafloor rocks, magnetic reversals caused an alternating

Archaeopteryx. Source: **The Natural History Museum, London.**

"striping," which was explainable only if the rocks had arisen from the ridge as magma, hardened, and moved outward, each carrying within it a "compass." Further evidence cascaded in, and now continental drift is considered a major feature of the Earth's continuing metabolism. Wegener was vindicated, albeit not in his lifetime.

Justice was dealt more swiftly when, not long ago, British astronomer Fred Hoyle put on his high-status armor and entered the arena of paleontology. Hoyle, who had made many important contributions to astronomy, took a skeptical look at *Archaeopteryx*, perhaps the single most important fossil creature in the scientific record. In all, there are but five specimens. *Archaeopteryx* means "ancient wing or feather." It was a small flying reptile, about the size of a crow, but had the remarkable feature of feathers, which are clearly imprinted in the limestone surrounding its fossils.

During the last century, when there was much speculation and discussion of Darwinian evolution, a major objection had to do with missing links. As one anti-Darwinist scoffed,

> We defy any one, from Mr. Darwin downwards, to show us the link between the fish and the man. Let them catch a mermaid, and they will find the missing link.

Archaeopteryx was just such a missing link, an ancestral species in between two great classes of animals, reptiles and birds. It was eloquently espoused by Darwin's foremost disciple of the time, Thomas Henry Huxley, and is still considered a prime example of Darwinian theory. Nonetheless, over the years, some have suggested that *Archaeopteryx*, known only from five fossils, is a hoax, and in 1985 they were joined by Hoyle, who inspected the fossil in the British Museum and found evidence of fraud.

He and an associate at University College Cardiff, astrophysicist Chandra Wickramasinghe, examined and photographed the fossil and declared that forgers had removed a layer of rock from around the reptilian bones and, using real and modern feathers, imprinted the fossil plumage. Part of the matrix, they also claimed, looked fake—rather like "flattened blobs of chewing gum." They suggested that the Häberlein family, who had discovered the first two *Archaeopteryx* specimens in a limestone quarry in Germany early in the nineteenth century, arranged the hoax to increase their value. The fossils were indeed sold—one to the Berlin Museum and one to the British Museum, where none other than Richard Owen (who had given the dinosaurs their name) was presiding. Hoyle and his associate implicated Owen in the hoax as well. Owen knew full well the fossil was a fake, Hoyle suggested, and bought it with the intention of exposing it and embarrassing Darwin. (In fact, once he obtained the specimen, Owen wrote the definitive description of *Archaeopteryx* and never made any attempt thereafter to say it was a hoax.)

As for Hoyle's 1985 claims of forgery, they were hard to reconcile with a similar fossil excavated as late as 1955, and another discovered in 1970 in a Netherlands museum, where it had sat unidentified and unnoticed since 1855. Even so, scientists at the British Museum undertook various tests of their *Archaeopteryx*. Among other things, they found tiny cracks in the fossil that extended through both matrix and feather imprint, making the method of hoaxing alleged by Hoyle impossible. The museum succinctly announced that Hoyle's claim was "codswallop," and the matter died away.

(There may have been more than merely a scientific agenda in at least Hoyle's colleague's mind. Previously, Wickramasinghe had testified on behalf of teaching creation science in an Arkansas trial on evolution in the schools. Perhaps he hoped to accomplish all these years later what he accused Richard Owen of wanting to accomplish—to poke a hole in evolutionary theory.)

Certainly Hoyle had jumped, rather foolishly it turned out, into a realm of which he had little understanding. What, then, of Luis Alvarez?

He was a giant in his own field of physics, and he commanded the respect due a giant. An experimental physicist, he was responsible for the discovery of many resonance particles (which have vanishingly brief lives and occur in high-energy nuclear collisions). During World War II he worked on microwave radar systems at MIT and then went to Los Alamos, where he suggested the technique used for triggering implosion-type A-bombs. He helped to develop the first proton linear accelerator, helped develop the use of radar for locating bombing targets, and had an important role in creating the liquid hydrogen bubble chamber used to detect subatomic particles. In addition to these achievements, he was aboard the *Enola Gay* when it took the atomic bomb to Hiroshima. He had a reputation for being loyal and helpful to his scientific co-workers, contributing much to their advancement up the ladder of success. Clearly this was a man who, when he spoke about the K-T boundary in 1980, commanded and deserved an attentive audience.

There was some reason as well to pause. Physics is a different world altogether from geology. In physics, typically, one can build a machine and devise an experiment to test a given hypothesis (both of which Alvarez was exceptionally good at), and then the experiment can be repeated by others to verify the results. Subsequent experiments can be conducted to test logical corollaries to the original hypothesis, and in due course—*quod erat demonstrandum*.

In geology, matters are not so clean, test procedures not so precise. All one has is data—and precious little of that—left from an "experiment" conducted tens to hundreds of millions of years ago. Not only can the experiment not be repeated, but a basic question is often, *What was the experiment?* What data is available can almost always be interpreted in a variety of ways, and only by surrounding the problem with more and different kinds of data can consensus be reached. Furthermore, that consensus can evaporate in the face of new data, either quickly or slowly as in the case of continental drift. Also, a geological consensus bears little or no relation to a proof in physics.

The trouble began almost immediately upon publication of the initial article. Many physical scientists applauded the Alvarez hypothesis; most Earth scientists found it absurd. Before we look at the early dissenters and what befell them, however, it is time to look a bit more closely at the putative culprit and its victims—that is, meteorites, and the dinosaurs and marine organisms that Alvarez said were done in so catastrophic a fashion.

Along the way we shall need a geologic timetable: a listing, so to speak, of Earth history with faunal assemblages, names, and dates. From early in the history of geology, it has been known that there were three distinct eras with evident and abundant life. Each had its own distinctive faunal assemblage, and each was separated from the next by a massive faunal turnover—what we now refer to as a mass extinction event. The first era is known as the Paleozoic (meaning, "ancient life"); the fauna was mainly aquatic, such as the trilobites seen in nature stores everywhere. The second is known as the Mesozoic ("middle life"); it is referred to as the age of the reptiles, or dinosaurs. The third is the Cenozoic ("modern life") and is often referred to as the age of the mammals. The Paleozoic and Mesozoic eras are separated by as the Permian-Triassic boundary and the Mesozoic and Cenozoic by the Cretaceous-Tertiary boundary. The whole sequence of eras is known, in geologic jargon, as the Phanerozoic ("evident life") eon. It is now known that lower forms of life existed in the eons preceding the Phanerozoic, but abundant life-forms began with the Paleozoic.

It was not long before these three recognizable eras were filled in with subdivisions, known as periods, epochs, and ages, in decreasing order of time duration. Each has its own distinctive faunal assemblage, and some, but by no means all, are separated by extinction boundaries. The following two figures, from a 1983 compilation, show these divisions as we know them today for the entire Phanerozoic eon and in more detail for the ages around the K-T boundary.

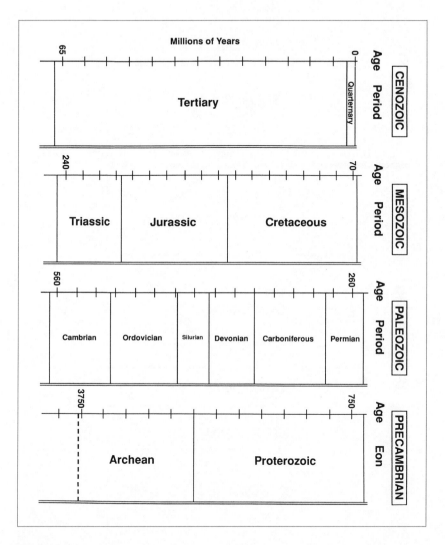

1983 geologic timescale. Ages are in millions of years. *Source:* **Decade of North American Geology, Geological Society of America.**

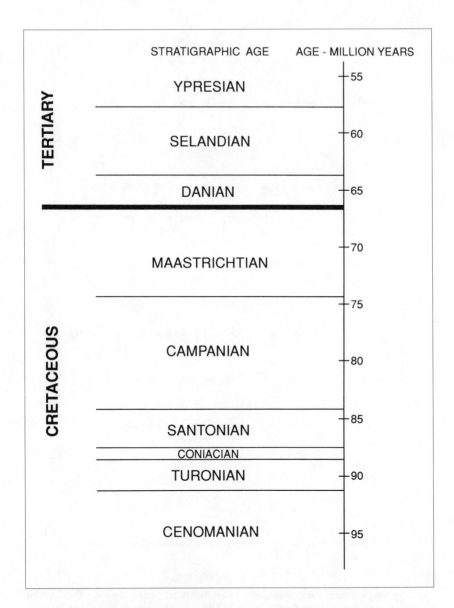

Detail of 1983 geologic timescale from the Cretaceous to the Tertiary period. *Source:* **Decade of North America Geology, Geological Society of America.**

TWO

Meteorites and Comets

JUST AS THE FBI MAINTAINS AND UPDATES personality profiles and other useful information about serial killers, arsonists, and the like, it is a good idea to do a background check on meteorites in general, to look into their origins, their characters, and their modus operandi, as it were. Much of what we hear about meteorites these days, like crime news, is confused and confusing, particularly when it comes to the danger they posed to ancient life, not to mention to civilization as we know it today. At the risk of sounding a bit schoolmasterish, and by no means downplaying the dramatic qualities of these alien and menacing visitors, it is important to get our terms straight.

A great deal of interplanetary debris was left over from the coalescence of material into the solar system. This debris ranges from microscopic specks of dust to chunks the size of large mountains and even greater. When a piece of such debris enters the Earth's atmosphere, friction soon causes it to grow very hot and to glow, at which point it is classified as a *meteor*. If the chunk is large enough to survive the abrasive inferno it creates by becoming a meteor and strikes the Earth, at that time (and not before) it is properly considered a *meteorite*. Very few meteors are large enough to become meteorites, but it was presumably not always so.

All it takes is a pair of binoculars to see that meteorites have made the Moon a pretty inhospitable place—and by implication the Earth, the major partner in this "two-planet" system. The Moon's surface, particularly its highlands, is pocked by craters great and small. There are craters within craters, bespeaking devastating events about which astronomers argue. Clavius, for example, the largest crater on the visible side of the Moon, is 230 kilometers (about 150 miles) in diameter and is itself pockmarked by more than a half-dozen good-sized craters. Some of the larger ones were probably caused by volcanic eruptions, but unquestionably the Moon was subject to a titanic battering by meteorites large and small for a very long time, especially during the solar system's formative years.

Since there never was any water or any atmosphere to speak of on the Moon (hence no *meteors*) and probably few tectonic movements of the crust and little erosion, virtually all of these pockmarks, little and gigantic, remain. On Earth, blessed as it is by large amounts of water and far more activity, the results of such bombardment from without and volcanic activity from within have been largely wiped away. Certainly both bodies are still struck by debris, but far less so than formerly, now that the Sun's gravity (among other physical forces) has swept up the neighborhood, so to speak, and imposed a more orderly existence in this tiny region of the universe.

But interlopers do come our way. The most recent notable interloper into our neighborhood in space was the comet Shoemaker-Levy (about which more will be said).

For most of human history comets have been regarded as bad omens. More than three thousand years ago, in 1059 B.C., the Chinese took note of what was probably Halley's comet, and its arrival was certainly documented in A.D. 88, as well as each of its subsequent twenty-seven returns in the last two millennia. In 1066 it showed up with a highly visible tail at the time of the Battle of

The return of Halley's comet in 1066, depicted in the famous Bayeux tapestry. *Source:* Tapisserie Bayeux, Bayeux, France.

Hastings. On the Bayeux tapestry, which records that great moment in history, King Harold II of England is shown sitting in obvious discomfort on his throne while alarmed subalterns point to the comet overhead—a bad omen for Harold, but perhaps taken otherwise by William the Conqueror. On a later return in 1456 (and of course no one then knew it was the same comet), Pope Calixtus III branded it "an agent of the Devil."

Not much fuss was made about Halley's comet when it dutifully showed up in 1835, but its next appearance in 1910 provided a spectacularly vivid celestial show, having a long tail that stood straight up, as one observer put it, "like the rays of a very powerful searchlight" shining from the horizon to the "roof of the heavens." In much of the world people were terrified; in Europe and America it was an occasion for great galas. A "comet cocktail" promised that an imbiber would see stars; comets appeared in advertisements for soap, coffee, tea, yeast, light bulbs, fountain pens, chewing tobacco, worm pills, machine tools, soft drinks, beer, and champagne (proving that *Jurassic Park* dinosaur mugs at McDonald's is not an original marketing concept). Doomsayers promised the end of the world: cyanogen, a deadly gas and minor volatile constituent of comets, would poison the Earth as it passed through the comet's toxic tail. A Haitian entrepreneur peddled pills to combat comet-sickness, and a flood in Italy soon was laid at its door.

If this all seems quaint now, the comet named for Lubos Kahoutek, predicted to make a bright appearance in the night skies of 1973, caused an eerily similar hoopla. NASA grabbed the wings of happy chance, adding Kahoutek's imminent arrival to the list of reasons to spend large amounts of money on the Spacelab program. Airlines promoted special charter flights; T-shirts abounded; the media covered doomsayers; and even some astronomers who should have known better made dire predictions. The only untoward event finally associated with comet Kahoutek was that, entertainmentwise, it proved something of a dud, being nearly invisible.

Far from an entertainment dud, thanks in substantial part to NASA's photographs from the multibillion-dollar and flawed but functioning Hubble Space Telescope, was the stream of fragments of comet Shoemaker-Levy that plummeted into the planet Jupiter in July 1994. The largest of the planets, Jupiter's diameter is 138,000 kilometers (about 86,000 miles) or eleven times the Earth's. (Unlike the Earth, it consists mainly of hydrogen—a hydrogen gas atmosphere about one thousand kilometers thick surrounding a region of liquid hydrogen, inside which is liquid metallic hydrogen. There may be an Earth-sized rocky core at the center.) The biggest comet fragments that struck the Jovian atmosphere were three kilometers in diameter and produced Earth-sized blemishes—spectacular and titanic events, but not unexpected given the volatile nature of the Jovian atmosphere.

We know a great deal about these nomads in space. First of all, it is fair to say that comets are less important than they look. Mostly they are what might be called special light effects.

A comet is primarily a nucleus made up of small particles of interstellar dust (accounting for 20 percent of the nuclear mass) along with frozen gases like methane and ammonia, and water. These cores, dubbed "dirty snowballs" by one astronomer, range from maybe a hundred meters across to a few kilometers. As one of these snowballs nears the Sun, its outer layers of ice melt, releasing some of its interstellar dust and gases to form the brilliant "head" or *coma*. The coma glows in part from its own luminescence and in part from reflected sunlight—and it can be enormous. The coma of the Great Comet of 1811 was reckoned to be one million kilometers across, larger than the Sun itself. Most comas are far smaller, and to achieve the status of crowd-pleaser, a comet really needs not just a notable coma but a visible tail.

The tail is made up of both gas and dust trailing the nucleus, but as was noted about Halley's in 1910, the tail doesn't follow directly; instead it points up, directly away from the Sun. Once a comet is near enough to be "lit" by the Sun, it comes within the influence of the solar wind, a continuing stream of ionized particles that the Sun gives off in all directions. Locally, the solar wind is a major player in the Northern Lights, more properly called the aurora borealis, but it also extends far enough out in space to push a comet's highly tenuous tail of gas and dust straight away from the Sun. (The dustiest part of the tail, it has been noted, may curve somewhat, a compromise between the direction of the comet's motion and the influence of the solar wind.)

So far, some twenty comets have been observed on at least two periodic appearances, and astronomers have plotted their orbits with great accuracy, in the process finding that there are three typical comet orbits. *Short period* comets swing around the Sun (the *perihelion* of their orbit) and whiz off to a farthest point (*aphelion*) roughly as distant as the orbit of planet Jupiter. *Long period* comets have their farthest points in Neptune's orbital neighborhood. Then there are *parabolic* comets. They move along a parabolic path, arriving and departing the Sun's neighborhood in a curve so great that they don't seem to be in an orbit at all, and so far as anyone knows, no parabolic comet has ever been spotted twice. A Dutch astronomer, Jan Hendrik Oort, proposed that they *are* in orbits, but vast ones, with their aphelia located in an enormous frozen cloud a light year away from the Sun—a cloud that remains only hypothetical. The theory is

The great comet of 1843 as seen over Paris. It developed one of the longest tails on record. *Source:* **Guillemin 1875.**

that all the comets we see—parabolic, long period, and short period—originate in Oort's cloud and are driven into the gravitational arms of the Sun by the perturbance of a star or some other cause. Some of these are further diverted by Neptune or Jupiter and are doomed to be exiles in the solar system.

Comets are relatively fragile objects for all their marathon journeying, and there is evidence that in the long course of orbiting the Sun, a comet's nucleus wastes away until it become all tail—in essence, a long, tenuous dust stream that extends along all or most of the former comet's orbit. When the Earth's atmosphere passes through such an old attenuated tail, some of the dust becomes a meteor shower—a beautiful array of shooting stars like the annual Perseid showers in August (which are less hazardous to human life and civilization than a Fourth of July sparkler).

A parabolic comet will take millions of years to return to the solar system (if indeed it ever does); at the other extreme is Encke's comet, which metronomically swings through every three and a third years. In all, there could be a hundred *billion* comets in

Oort's cloud, with very few ever likely to be jostled out of it. If they were all combined into one big snowball, however, the entire mass would be less than that of the Earth. Indeed, so small in mass and density is Halley's comet (or any other) that if the Earth were to pass directly through the glowing head (as opposed to the nucleus), the result would probably be nothing more than a bright meteor shower. A comet tail's density is equivalent to the vacuum produced in a decent laboratory vacuum apparatus. A comet's nucleus might wreak a bit of havoc if it hit the Earth, but of the estimated 2,000 comets in the grip of the Sun's gravity, none come particularly close to Earth. Halley's nearest approach was 23 million kilometers (about 14 million miles) in 1910.

It would be injudicious to say that comets pose absolutely no threat to the Earth. But of far more concern are the ever-so-much-brawnier meteoroids, which many assume are wayward visitors from the asteroid belt, which lies between the planets Mars and Jupiter, between 240 and 800 million kilometers from the Sun. This ring of rubble never managed to coalesce into a planet when the solar system came into its own. Again, not to diminish the dramatic nature of these objects, it should be pointed out that *no one in all of human history* (or in the myths that come down to us from prehistory) *has been struck by a meteorite*—except a woman in Alabama who was hit in the arm when a meteorite plummeted through her roof in 1954. There are some 40,000 asteroids in the asteroid belt, and the vast proportion of them stay put out there beyond Mars. Still, at least four of them—Adonis, Apollo, Geographos, and Hermes—with diameters of a kilometer or less, approach to within about a million kilometers of the Earth.

Chunks of material with asteroidlike composition occasionally make it through the atmosphere and hit the ground as meteorites, typically the size of a baseball or smaller. Large ones are extremely rare; the largest recovered in North America was less than two meters across.

These interlopers have been known for millennia and puzzled over for just as long. For a long time their bright and loud entry into human affairs was no doubt taken as a raging sign of one or another deity's wrath, and more recently some who have found them lying improbably on the ground have thought they were stones carried upward in waterspouts and dropped by associated thunderstorms. An early scientist to suggest they were extraterrestrial in origin was German physicist Ernst Chladni in 1794, and he was met, not surprisingly, with scorn. A colleague retorted that reading Chladni's theory at first made him feel as if he had been hit on the head by one of these rocks. Another called Chladni one of those who "deny any world order, and do not realize how much they are to blame for all the evil in the moral world."

Shades of Wegener. Indeed, shades of Pope Calixtus III.

Not even Thomas Jefferson, with his fine scientific mind, gave much credence to the extraterrestrial hypothesis. On December 26, 1807, a meteorite fall occurred in Weston, Connecticut, particles of which were identified as meteoritic by two Yale geologists, who also spoke for their extraterrestrial origin. In what is probably the only example of a sitting U.S. president entering directly into a scientific debate, Jefferson commented that he "would rather believe that those two Yankee professors would lie than to believe that stones fell from heaven."

But fall they do, and in the past some very large ones have pounded the Earth, though as noted before, not many craters have survived the Earth's geological eraser. In recent years geologists have found scars of impact craters on Canada's Precambrian Shield, a vast tract of ancient rock—some of the scars date back more than a half-billion years. The largest of these is Manicouagan, with a diameter of 100 kilometers and an age of 214 million years. (Manicouagan is *half* the size of the K-T crater that Alvarez suggested; and it is to be noted that *no* mass extinction event is associated with the Manicouagan impact.) In all, there are around a hundred known or probable meteorite craters of significant size on Earth.

Meteor Crater, the most unquestioned impact of any size, is a well-known tourist attraction just south of what is now Route 40 between Flagstaff and Winslow, Arizona. Called Barringer Crater as well, it is the most recent, having occurred about 25,000 years ago (before any reliable evidence of human habitation in this hemisphere), and it is the most thoroughly studied. Indeed, it was during a period stretching from the late nineteenth into the twentieth century that students of this crater determined that the meteoritic impact was an actual, bona-fide geological process. Once called Coon Butte, it is a low-lying feature on the landscape about one kilometer across, with a rim that reaches fifty meters and an interior depression some two hundred meters deep. People poking around on the floor of the depression would find fragments of oddly formed iron weighing up to a kilogram or so, and other smaller hunks were scattered around the landscape in nearby Diablo Canyon. A few chunks from within the rim weighed as much as a few hundred kilograms, which caught the attention of Grove K. Gilbert, a founding member of the U.S. Geological Survey and its chief geologist from 1879 until his death in 1918. If it were a meteorite that struck at relatively low speed, Gilbert assumed logically but (it turned out) incorrectly, the object would have scoured

Aerial view of Meteor Crater, Arizona. *Source:* **United States Geological Survey, Denver.**

out the crater and buried itself beneath the crater floor. But a magnetic survey across the crater floor showed no sign of it, so Gilbert concluded it had been formed in a huge steam explosion of volcanic origin, and the meteoritic fragments were incidental.

In 1902 a lawyer and mining engineer named Daniel Moreau Barringer from Tucson heard about all this and, assuming the crater was meteoritic in origin, computed that a body of ten million tons of nickel, platinum, and iron lay underneath it, a nest egg to be dug up worth a half-billion dollars. So he staked a claim to the crater and drilled on and off until 1928, never finding the mother lode or any other kind of lode. Even so, Barringer remained convinced it *was* a meteorite crater and was always annoyed that the Survey would not acknowledge his interpretation as valid. While Gilbert probably agreed, he remained silent, and the Survey continued officially to espouse a volcanic origin for several more years.

Today it is universally acknowledged to be the result of a meteorite, and the frustrating nonexistence of the ore body is now well understood as part of what happens in the formation of a classic meteorite crater. Gilbert and Barringer both made a wrong assumption, ascribing a relatively low velocity to the incoming projectile. The kinetic energy of a moving body is proportional to its mass and also to the square of its velocity. If the velocity is raised ten times, the mass needed to achieve the same kinetic energy (that is, in this case, to produce a depression the size of Meteor Crater) is one hundred times less. Current thinking is that the object that excavated Coon Butte came in a little faster than an artillery shell and largely vaporized on impact, leaving only scattered fragments, some of which were forced deep down into the ground below the crater floor. Rather than the mouthwatering ten million tons of highly salable metal envisioned by Barringer, the object is thought today to have been only 300,000 tons and only 50 meters across.

In what is now considered the textbook manner, the impact created an enormous shock wave that spread through the surrounding rock at a tremendous speed, vaporizing some rock, melting some, crushing, fragmenting, and excavating others. Some of this material, including some meteorite fragments, would have been thrown out at low angles into the surrounding countryside—these are called *throwout deposits*. Some would have been thrown more directly upward, falling back into the crater as rock fragments called *breccia*. In impacts of this sort, temperatures can reach 3600°C, turning the rock to liquid. This liquid rock may escape in hot jets, cooling into curiously shaped objects called tektites, which sometimes are found in the form of buttons and dumbbells. Some of the rock that metamorphoses in the extreme pressure takes on different kinds of crystal forms, including unusual crystalline markings, which we will discuss in due course.

More complex craters have been found (though none so pristine as Arizona's) that are multiringed, with far more complicated rupturing, domelike peaks in the center, and diameter-to-depth ratios of 10:1—that is, far shallower. Some have speculated that these shallower craters are caused by lower-density comet nuclei whose energy would be released nearer the surface, but no one knows for sure. (It is interesting to note that for an impact great enough to cause the kind of subsequent phenomena proposed by the Alvarez hypothesis, the crater's diameter-to-depth ratio would have to be far more severe than that of Meteor Crater—something on the order of 15:1 or 20:1.)

In historical time a huge event occurred that was almost certainly the result of some extraterrestrial visitation—but of what remains totally ambiguous. It occurred in the Tunguska River region of Siberia some 600 kilometers north of Vladivostok on June 30, 1908. That morning, passengers on the Trans-Siberian Railroad were stunned to see a meteor as bright as the Sun race across the sky from south to north and disappear beyond the horizon. Immediately, they felt a violent blast of air. Later it was learned that the blast had been felt over a distance of 80 kilometers and flattened forests like so many matchsticks, in a radial pattern extending 30 kilometers from the center. It was a spectacular event, to be sure, but nothing like a major earthquake in destruction.

The culprit was assumed to be a meteorite, but expeditions to the remote site have never turned up any meteoritic particles. Yet something had screamed into the Earth's atmosphere at a shallow angle, something big enough to survive a long and fiery entry, and it apparently exploded about ten kilometers above the ground. The destruction on the ground, it was determined right away, was not from any solid object but rather the blast wave of an airborne explosion.

So what was it? Nobody knows for sure. Some scientists have suggested it was a small comet, maybe a fragment of that regular visitor, Encke's comet; others think it was a chunk of asteroid. It could have been only a few tons in mass or as much as 50,000 tons—meaning an object a few tens of meters across. At the worst, this single known interplanetary havoc-wreaker in modern times was a relatively small object that had minimal effects.

Even so, one would not have enjoyed standing near the Tunguska River that day. Since the Tunguska "event" (whatever it was) and particularly since the Alvarez hypothesis and the 1994 battering of Jupiter, astronomers have called for large amounts of funding for sentinel duty. In particular, they have mounted a search for another class of interloper—near-Earth asteroids (NEAs), swarms of objects that share the inner solar system with us. Astronomers have catalogued about two hundred of these odd-shaped bodies, ranging in length from ten meters to a few kilometers. About 10 percent of them appear to cross the Earth's orbit every few thousand years, then circle back out to Martian realms. It has been suggested that this class of objects could have been irregularly dislodged from the asteroid belt or a separate asteroid belt, or perhaps from the remains of dead comets. Whatever they are, investigators say there is little risk to fret about since most of these objects are not big enough to make it through the Earth's atmosphere, the occasional alarmist newspaper story to the contrary notwithstanding.

What is the risk that the Earth will be struck in the near future by a meteorite big enough to cause severe damage? Of course, nobody knows for sure, the future being largely unknown and many potential meteorites being as yet unidentified. But various scientists have computed the odds. One estimate that appeared in the British science journal *Nature*, by two American researchers (Clark R. Chapman of Tucson's Planetary Research Institute and David

Morrison of NASA's Ames Research Center), said there is a one-in-ten-thousand chance that an asteroidal or cometary meteorite two kilometers wide will strike the Earth in the next hundred years.

The Earth, however, is itself a very big place, and where in its vast reaches (mostly ocean) a meteorite might strike is of obvious importance. In his book *Technological Risk* H. R. Lewis, a professor of physics at the University of California at Santa Barbara, addressed this question as one who has served on various federal government risk-assessment panels on defense and nuclear power matters. What are the odds of a large meteorite striking a U.S. city in the next ten years? he asked. There are forty metropolitan areas with populations of a million or more, he pointed out, comprising a total land area of 3,000 square miles. The United States as a whole contains three million square miles, so the odds of a meteorite hitting a big city are, to begin with, one in a thousand. Then one must factor in the time element. Lewis pointed to Meteor Crater, saying that if it hit 50,000 years ago, we can postulate such a strike every 100,000 years. This comes out to a chance of one in ten *million* that such a meteorite would hit an American city in the next decade—not something, he says, to lose much sleep over. (The odds are a bit less if the date commonly attributed by geologists to Meteor Crater—25,000 years—is factored in.)

And how should one compare this risk to that of, say, an earthquake? Apples and oranges, perhaps, but it is noteworthy that in 1976 the Chinese industrial city of Tangshan, with a population of one million, was reduced to rubble by an earthquake, and one-quarter of its population died. Between 1949 and 1976, the People's Republic of China reported that 27 million people died and 76 million more were injured as a result of a hundred earthquakes within its borders. Even so, even with such cataclysmic destruction, the course of civilization has not been radically altered, a risk that newspaper interviews with astronomers about potential meteorite strikes usually cite alarmingly.

But let us go on to see what the geologic record can tell us about meteorites and extinctions. The Alvarez hypothesis, as one might expect, has given rise to a number of studies of the biotic effects at known impact sites. They have shown no global extinction effects. There have not even been any *regional* extinction effects at these sites.

Next to Meteor Crater in Arizona, the Ries Crater in southern Germany has been the most thoroughly studied impact site. The Ries Crater has a diameter of 24 kilometers and an age of 15 million years. Kurt Hessig of the Bayerische Staatssammlung für Paläontologie in Munich found no abrupt change in the mammalian faunal distribution in nearby sedimentary sections at the time of the Ries event.

The Montagnais Crater is a buried impact structure, as determined from seismic reflection profiling and drilling, off the coast of Nova Scotia. It has a diameter of 45 kilometers and an age of 51 million years. Marie-Pierre Aubry of the Woods Hole Oceanographic Institution and her colleagues found no change in the microfossil distribution from wells at this site at the time of this impact.

Farther to the southeast, on the continental shelf off Virginia and New Jersey, Wylie Poag of the U.S. Geological Survey and colleagues have found, again by seismic reflection profiling and drilling, two additional impact structures in the sedimentary section. The Chesapeake Bay Crater is quite large, with a diameter of 85 kilometers; the adjacent Toms Canyon Crater has a diameter of 20 kilometers. Both are 35 million years old. But in this much-explored sedimentary section off the east coast of the United States, there is no regional extinction event corresponding to the 35-million-year age.

Finally, there is the Red Wing Creek Crater in North Dakota. It is a buried structure of nine kilometers diameter with an age of 200 million years and is presumed to be of impact origin. It is an oil-producing structure in the Williston Basin—the fractured rocks of the structure provide increased porosity to permit petroleum accumulation, a most unusual oil field. There are no recorded faunal disruptions of the structure's age in the Williston Basin sedimentary sequence.

Well, let's give those who consider impacts to be the causative agents for mass extinctions one more chance. The two major

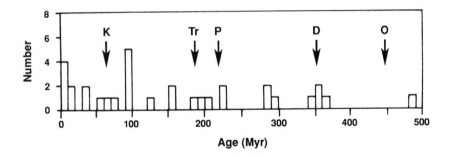

Distribution of ages for dated craters on the Earth with diameter greater than 10 kilometers. Arrows indicate the approximate times of the five major biological extinctions of the past 500 million years: End Cretaceous, or K-T; End Triassic; End Permian, or P-Tr; End Devonian; and End Ordovician. *Source:* **Weissman 1985.**

extinction events that separate the three major divisions in the geologic record occurred at the end of the Permian, the Permian-Triassic (P-Tr), and at the end of the Cretaceous, or K-T. Along with these two there are three other extinctions of substantial magnitude. They constitute the "big five" mass extinctions. These are the End Ordovician, End Devonian, and End Triassic. Is there any correlation between these extinction events and the impact record on Earth? The diagram by Paul Weissman of the Jet Propulsion Laboratory shows such a comparison. Quite simply, there is no obvious one-to-one relationship between known impacts and major biological extinctions.

So, there you are. Nothing plus nothing still equals nothing.

THREE
A Brief History of Dinosaurdom

WITHIN THE ARENA OF SOCIOLOGY, WHICH ROBERT
Jastrow's one-time colleague did not count among the hierarchy of
sciences, the subdiscipline of criminology is seeing yet a further subdiscipline emerge: victimology. What personal characteristics, it is
asked, make someone likely to be a victim? Some of them have to
do with posture and gait, as in the way members of street gangs
react to one another. Anyone with a wimpy demeanor or gait is
going to be beaten on in a kind of pecking-order syndrome, whereas
someone who strides purposefully and defiantly about, staring potential competitors straight in the eye, may avoid physical confron-
tation and even earn a certain amount of "respect."

It is not completely far-fetched to bring this up in a profile of
dinosaurs as victims, since their posture and gait are among the
biological features that truly distinguished them from the other reptilian creatures with whom they shared the planet for 150 million
years. Not only that, it might have been matters associated with this
gait that allowed them to be done in at the end of the Cretaceous.

The Mesozoic era has been aptly called the Age of Reptiles, but the dinosaurs were a long time coming. Whatever was the progenitor of the dinosaurs, it lived at a time when the world was, in essence, already full of reptilian types filling various niches. This Triassic period saw the beginnings not only of the older amphibian and reptilian lineages but of frogs, crocodilians, lizards, and turtles, not to mention the earliest mammals.

Had all these reptilian types died out at the end of the Cretaceous, we might look back with the most awe at the turtles. For here, having evolved out of some primitive lizardy creature, is a wholly unlikely beast. It has a beak, not teeth; a hard shell attached to its ribs, meaning its backbone is no longer flexible but rigid; and another shell covering its stomach. A beast given to life both on land and in the water, it propels itself not by swishing its vertebrae back and forth in the accepted manner of fish, but by paddling with four feet. These comparatively oafish animals nonetheless became the only reptiles known to be capable of migrating for thousands of miles. Compared to the mystery of turtle evolution, the springing of crocodiles, snakes, and even dinosaurs from some primitive lizardy stock seems pretty straightforward.

Looked at closely, the appearance of dinosaurs can seem something like a happy accident. Paleontologists have found several candidates for dinosaur progenitor, but the entire matter is still murky. Whatever it was, the progenitor had to have been at first a very minor player ecologically in terms of numbers, making a living in a world already abundantly full of reptiles, most of them bigger than itself. It may have preyed on reptilian young, or on protomammals (themselves ecologically insignificant), or maybe even on both of the above plus the occasional dragonfly. Nobody knows. But a mere ten million years later, about 215 million years ago, dinosaurs had become the dominant land animals. The chances are that their success lay in their gait—which was bipedal—and a bipedal gait, in turn, is associated with the pelvis and some other important matters that people would not immediately associate with the pelvis. But first, a more general question: What was the Mesozoic like in terms of climate and other environmental features?

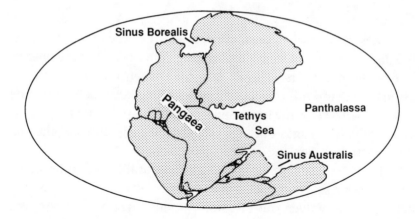

The ancient land mass of Pangaea as it may have looked 200 million years ago. *Source:* **Uyeda 1971.**

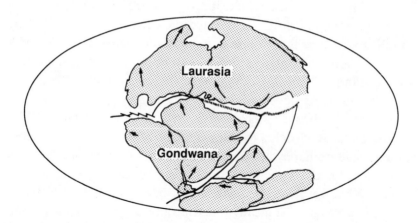

World geography 180 million years ago, during the Jurassic period. *Source:* **Uyeda 1971.**

To begin with, near the beginning of the Mesozoic there was one huge continent we now call Pangaea, surrounded by a global ocean we call Panthalassa. On the continent's east was a partly enclosed sea, called Tethys, which would one day become the Mediterranean. There was no Atlantic Ocean. By 180 million years ago, the planet's continents were barely recognizable. The ancient landmass had split into two giant forms: Laurasia (consisting of present-

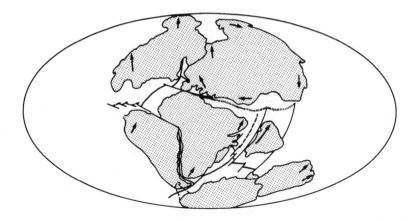

World geography 135 million years ago, near the beginning of the Cretaceous period. *Source:* **Uyeda 1971.**

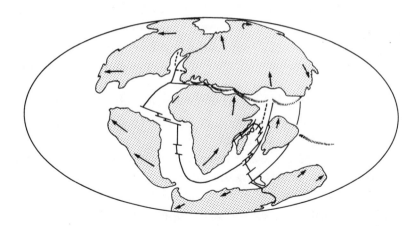

World geography 65 million years ago, at the end of the Cretaceous period. *Source:* **Uyeda 1971.**

day North America and Eurasia) and Gondwana (consisting of South America and Africa). In addition, a huge body that would eventually become Australia and Antarctica had broken off from Gondwana, as had the subcontinent of India. Launched on a northeasterly course, the Indian subcontinent would ultimately plow into Asia, creating the vast wrinkle in the Earth called the Himalayas. By the beginning of the Cretaceous period, South America had begun to split off from Africa, creating the early Atlantic Ocean, and by the end of the Cretaceous, some 65 million years later, the continents were beginning to resemble what we see today, though North America would later split off from Eurasia, as would Antarctica from Australia. One might think that, all this titanic activity would have thrown the climate into a long series of changes as well—a climate rollercoaster. And that did happen, as we know from studies of the oxygen content of oceanic sediments.

Among its numerous roles on Earth, oxygen is a constituent of the skeletal parts of carbonate-bearing plankton, the microscopic creatures that drift around the oceans. These planktonic "shells" form the bulk of the deep-ocean sediment into which we can now drill, bringing up cores that take us back through long sequences of time. Happily for us (and no doubt for the plankton as well), two stable forms of the element oxygen occur in nature. The most common form, or isotope, has an atomic weight of 16, while the less common isotope is 18. Both of these isotopes react to form such compounds as calcium carbonate, but they do so at different *rates*. They also do so when seawater evaporates from the ocean surface. So when seawater and a carbonate that is dissolved in it react, the ratio of oxygen-18 to oxygen-16 will be higher in the carbonate plankton than in the surrounding seawater. Not only that, but the lower the water temperature, the higher the ratio of oxygen-18 to oxygen-16. In this discrepancy we have a reliable paleothermometer for measuring the temperature of ancient seas. Even more helpful, ocean surface temperatures are a good proxy for atmospheric temperatures.

It has been found that mean surface temperatures through most of the Mesozoic were between ten and twenty degrees Fahrenheit warmer than temperatures today. Then, starting about 120 million years ago, these temperatures began an orderly decline to those of today. The world evidently luxuriated in a long benign period of steady greenhouse warmth; 120 million years ago there evidently was six times as much carbon dioxide in the atmosphere as now. The decrease in temperature was gradual, and it showed up mostly in temperature gradients from the polar regions to the equator—but throughout these times and into our own, those

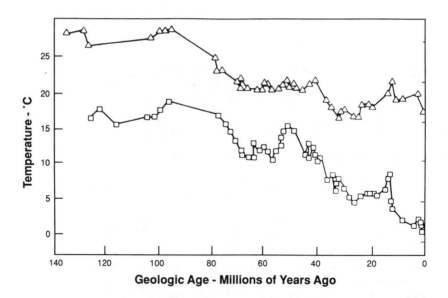

Surface and bottom temperatures, inferred from oxygen isotope measurements, for low latitudes near the equator in the Pacific Ocean versus geologic time. Surface temperatures are shown as triangles and bottom temperatures as squares. *Source:* **Savin 1982.**

most temperature-sensitive land animals, the amphibians, made out well enough in most parts of the world. Into this kindly climate and reptile-rich environment, the dinosaurs emerged—probably from a single progenitor—and soon differentiated into two basic lineages, called *orders*, which paleontologists distinguish by the hips or pelvises. A pelvis has an upper bone, the *ilium*; a forward bone, the *pubis*; and a backmost bone, the *ischium*. In one line of dinosaurs, the pubis is swung downward and backward so that it is parallel to the ischium, and since this is much like the pelvic arrangement of birds, these dinosaurs are called *ornithischian.* The other order is called *saurischian* because the hips are lizardlike, with the pubis forward of the ischium. In both cases, the configuration of the hips suggests that the hind legs were more vertical than the splayed-out legs typical of lizards and turtles, and this almost certainly made dinosaurs more agile than other reptiles.

The ornithischian lineage, over time, would prove highly creative, producing a fantastic array of plant-eaters in myriad shapes

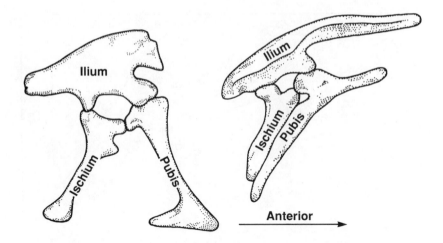

Basis for the division of dinosaurs into two orders. On the left is the arrangement of pelvic bones in the saurischians; on the right is the arrangement of bones in the ornithiscians. *Source:* **Levin 1991.**

and sizes, including the familiar *Stegosaurus*, which reverted to walking on all four legs, and the goofy-looking duckbilled dinosaurs.

The saurischians would produce both plant-eaters and carnivores—the enormous *Apatosaurus*, a well-known plant-eater (which also reverted to walking on four legs), and the bipedal *Tyrannosaurus*, the most appalling carnivore in world history. What is of particular interest to us is that the vertical hind legs of virtually all dinosaurs, a function of their pelvises, allowed them to achieve great size. To be sure, there were many small dinosaurs, but most species weighed a ton or more. (By comparison, a typical horse weighs a bit more than a half-ton.) Being big may have had a great deal to do with their becoming, toward the end of the Triassic, the boss animal and with their continuing hegemony through the Jurassic and Cretaceous. But their competitive edge is probably also associated with their upright gait.

The upright gait itself suggests that these dinosaurs were agile *and* capable of extended periods of activity. A typical reptile cannot run for very long; one reason is that it depends on a fermentive metabolism that creates a rapid buildup of lactic acid, after which it needs to rest for a prolonged period. On the other hand, dinosaurs may well have developed early on an oxydative metabolism that would have permitted the extended activity that their upright posture suggests. Whether this metabolism was like that of today's mammals and birds—that is, whether they were warm-blooded—is still a matter of great debate, one not easily resolved. If they were warm-blooded, they could have raced around the landscape

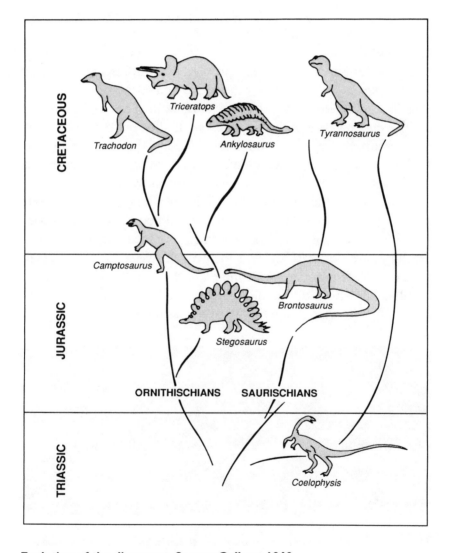

Evolution of the dinosaurs. *Source:* **Colbert 1969.**

whenever they pleased; if cold-blooded, they would have had to wait till the sun warmed them up to get under way. And they may have been different from both mammals and birds on the one metabolic hand and from today's reptiles on the other; fossilization simply does not preserve the soft tissues that would be helpful in making such a determination.

Among reptiles, the orders of lizards, snakes, turtles, and crocodiles survived beyond the Cretaceous and into modern times. The two orders of dinosaurs (the saurischians and the ornithischians) did not survive, nor did a number of other reptilian orders, mostly marine. Fishlike reptiles, the order of *ichthyosaurs*, dating back to the late Triassic, were prevalent in the Jurassic, but all died out before late Cretaceous times. They were reminiscent of today's porpoises, with tails that served as the main swimming organ. Another marine order was the *plesiosaurs*, ribbonlike reptiles, which are known into the late Cretaceous. Fitting the niche of today's killer whales, the plesiosaurs had long necks, short tails, and four large limbs in the form of paddles, which no doubt served as their means of locomotion. The closest thing to a plesiosaur today is the ever-elusive and always-hoped-for Loch Ness monster. Then there are the *placodonts*, or flat-teeth reptiles. They lived during the Triassic. Their limbs were short and powerful, and evidently they walked across the sea floor feeding on

Finely preserved fossil of *Ichthyosaurus acutirostris*. *Source:* **Natural History Museum, London.**

clams and the like. Finally, there are the *mosasaurs*, or reptiles from the Meuse. They evolved and lived during the late Cretaceous, with the tail as the main swimming organ. They were quite large, 15 to 30 feet long, had strong jaws, and must have been savage predators. They are similar to modern monitor lizards.

The Mesozoic air was the province of the *pterosaurs*, the flying reptiles or "dragons of the air" as they have been called, that were contemporaries of the dinosaurs throughout their time on Earth. They were nearly ubiquitous, being known from Asia, Europe, North America, and Australia, and they proliferated into many species. But as fliers, they were delicate in bone structure, with minimal chances of being preserved as fossils. Most of them have been found in marine deposits, suggesting that they lived lives similar to modern sea birds. The lack of fossil remains makes it difficult to track their extinction, but it is believed that by the late Cretaceous there were only a few species still flying.

The last refuge for the dinosaurs was apparently western North America, the only region in the world where abundant dinosaur remains have been found in a more or less continuous sequence extending through late Cretaceous time (the Campanian and Maastrichtian ages). Elsewhere the record is more scattered and less complete, but it does suggest that dinosaurs disappeared outside North America well before the Cretaceous came to an end. In South America and Asia, the last dinosaurs are Campanian to middle Maastrichtian in age. In northern Europe, only one and maybe two dinosaur species lasted into the Maastrichtian. And in North America, the dinosaur demise began as a gradual process, about seven million years before the K-T transition, accelerating in rate during the last 300,000 years. The last two known dinosaur species were the herbivore *Triceratops* and the carnivore *Tyrannosaurus*.

It is little wonder, then, that paleontologists took the Alvarez hypothesis, with its suggestion of an abrupt "lights out for the dinosaurs," as probably the wrong answer for the wrong question. What about the other reptile orders, they asked, that also disappeared before or at the K-T transition?

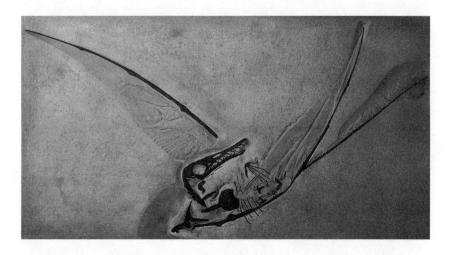

A fossil Jurassic pterosaur. *Source:* Natural History Museum, London.

Depiction of *Triceratops*, last of the herbivorous dinosaurs, and *Tyrannosaurus*, last of the carnivorous dinosaurs. *Source:* Field Museum of Natural History, Chicago.

The pterosaurs were present throughout Cretaceous times, but they began a marked decline some 40 million years before the K-T boundary, in Albian time. (Ten million years after their decline began in earnest, it is interesting to note, there was a large increase in bird species. Some scientists aver that since birds arose from dinosaurian progenitors, the dinosaurs *did not* go extinct, but this is basically quibbling.) Among other orders, the ichthyosaurs were gone by Campanian time, about 15 million years before the K-T boundary, and plesiosaurs diversified richly in late Jurassic and early Cretaceous times but were gone before the end of the Cretaceous, as were also the mosasaurs.

When it comes to the dinosaurs and related charismatic megafauna of the era, whatever did happen at the K-T transition appears to have been more a capstone to events that were already well under way. The truly vast extinctions of this time took place among microscopic sea organisms and marine shellfish—accounting for the disappearance of nearly 50 percent of all the world's species—and the Alvarez hypothesis purported to explain them, too.

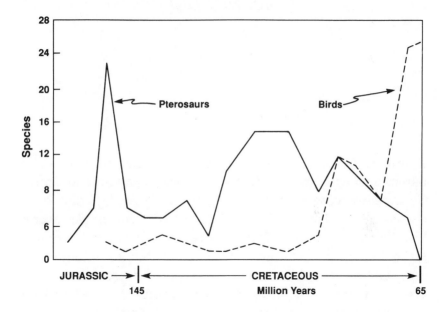

Diversity of pterosaurs and birds in the late Jurassic and Cretaceous periods. *Source:* Unwin 1987.

FOUR

Paleonecrology

WHEN THE MATTER OF DINOSAUR EXTINCTION LAPSED into its dilettante phase in the 1920s, it was in part because professional paleontologists had found an especially fertile field of study elsewhere—namely, in the fossil record of innumerable invertebrate marine organisms. This record extends over nearly unimaginably long periods of time. In spite of the notoriety of dinosaurs, especially in the public mind, it was among these marine organisms that the most devastating extinctions appeared to have taken place during late Cretaceous and K-T times.

If the dinosaurs were kings of the land, the *ammonites* were kings of the marine invertebrate realm. These mollusks lived in coiled, chambered shells not wholly unlike snails, and were the ecological counterparts of many shallow-water shellfish of today. It is not that unusual to find fossilized remains of these creatures in the open rock while hiking in a mountainous region—a sign that the highlands were once under a large sea. Even more plentiful than ammonites was the ocean's *plankton*—microscopic plants

A Middle Jurassic ammonite. *Source:* Natural History Museum, London.

and animals—represented in the fossil record by *foraminifera* (or microfossils) and *coccolithophoridae* (or nannofossils).

In 1980, when the Alvarez hypothesis exploded on the scene as a single, all-encompassing explanation for the K-T extinction events, paleontologists already had their own ideas about what had transpired among these marine invertebrates. Indeed, these very creatures had permitted scientists to pin down with increasing accuracy the actual timing of events that have taken place over the vast stretches of geologic time since life appeared on the planet. In their comings and goings as life-forms, marine invertebrates have left a calendrical record on the sea floors that could be read as soon as scientists had figured out how. In 1980, in other words, paleontologists already knew plenty about extinctions, including the mass extinctions that in a sense have punctuated geologic time.

As good a place as any to begin is back in the early 1800s, when geology was coming into its own as a science—amid, not surprisingly, a good deal of controversy. One of the controversies concerned the age of the Earth itself. It pitted geologists against an eminent physicist of the era—William Thomson, better known by his peerage title of Lord Kelvin—and both of these sides against the traditional view of the matter widely held by the church.

Biblical analysis—particularly of the Book of Genesis—had provided a wondrously precise date for the creation of Earth, mankind, and all other living things. In 1694, James Ussher, Episcopal archbishop of Armagh and primate of Ireland, carefully tracked all the biblical begats and concluded that Creation took place at exactly 9:00 A.M. on October 26, 4004 B.C. This astonishing "fact" might well have passed unnoticed, but for an unknown scribe who included it as a marginal reference note in the King James version of the Bible. It didn't take long for the date to become doctrine and for anyone who assumed another date, especially an earlier one, to be branded a heretic.

Lord Kelvin.
Source: Annan, Glasgow.

In the early eighteenth century, however, many geologists held a view that could be called *unlimited time.* Based on the fossil record that had come to light and from the many sequences that could be studied in sedimentary strata, they could only postulate an unlimited time for the evolution of all the varied species and for the succession of events—sedimentation, erosion, and resedimentation—that were obvious in the rocks.

In the latter part of the nineteenth century, Lord Kelvin challenged not only the geologists' vague notion of unlimited time but their strict adherence to a geologic principle called *uniformitarianism.* This principle said that the present is key to the past, that what is going on today has been going on in much the same manner throughout all of time. Kelvin was seeking a correct estimate for the age of the Earth, and though he got it wrong, in the process he made a substantial contribution to the field of geology.

He set about determining the Earth's age by means of some straightforward calculations, taking as his starting point the time when the outer portion of the Earth had solidified from a molten state. He then made a simple calculation from heat conduction theory: all he needed was the variation of temperature as a function of depth beneath the Earth's surface (its *thermal gradient*); the thermal conductivity of its rocks; and its initial temperature. The temperature gradient was known from measurements in deep mines; Kelvin determined the thermal conductivity of rocks in his laboratory; and he made a reasonable estimate for the temperature of molten rocks. Based on these figures he calculated several possible ages for the Earth: a maximum of 400 million years, 100 million, 50 million, and (in a final calculation made in 1897) 24 million years.

Geologists found these dates implausible, in particular his final one of 24 million years, which simply didn't seem long enough to permit all the activity they could read in the rocks. But their own notions, based as they were on guesses about the time needed for evolution and erosion, were hardly quantitative. Nevertheless, they took up Kelvin's challenge. One of their principal spokesmen was Thomas Henry Huxley, the great defender of Darwinian evolution, a superior orator, and an excellent scientist in his own right. In 1869 he wrote in a backhanded manner:

I do not suppose that, at the present day any geologists would be found to maintain absolute Uniformitarianism, to deny that the rapidity of the rotation of the earth *may* be diminishing, that the sun *may* be waxing dim, or that the earth itself *may* be cooling. Most of us, I suspect, are Gallios, "who care for none of these things," being of the opinion that, true or fictitious, they have made no practical difference to the earth, during the period of which a record is preserved in stratified deposits.

Huxley went on to challenge Kelvin's estimate directly:

Mathematics may be compared to a mill of exquisite workmanship, which grinds you stuff of any degree of fineness; but, nevertheless, *what you get out depends on what you put in*; and as the grandest mill in the world will not

Thomas Henry Huxley.
Source: Vanity Fair.

extract wheat-flour from peascod, so pages of formulae will not get a definite result out of loose data.

In other words, garbage in, garbage out—in the less elegant phraseology of the computer age.

As it turned out, Kelvin's age estimates were erroneous precisely because of his assumptions—specifically, he omitted the factor of radioactive heating within the Earth, a phenomenon unknown at the time he made his calculations; it was not discovered until 1896 by French scientist Henri Becquerel. Credit for emphasizing radioactivity's importance to the Earth's heat balance goes to another eminent British physicist, Lord Rutherford, who first presented his ideas about it at a lecture in 1904 at the Royal Institution. As he later described the moment:

> I came into the room, which was half dark, and presently spotted Lord Kelvin in the audience and realized that I was in for trouble at the last part of the speech dealing with the age of the earth, where my views conflicted with his. To my relief, Kelvin fell fast asleep, but as I came to the important point, I saw the old bird sit up, open an eye and cock a baleful glance at me! Then a sudden inspiration came, and I said Lord Kelvin had limited the age of the earth, *provided no new source of heat was discovered.* That prophetic utterance refers to what we are now considering tonight, radium! Behold! The old boy beamed upon me.

Rutherford concluded:

> The discovery of the radio-active elements, which in their disintegration liberate enormous amounts of energy, thus increases the possible limit of the duration of life on this planet, and allows the time claimed by the geologist and biologist for the process of evolution.

Whether Kelvin beamed at Rutherford or not, he never did agree with Rutherford's findings—not an unusual occurrence in geology debates. As yet another eminent physicist, Max Planck, has written: "A new scientific truth does not triumph by convincing its opponents and making them see the light, but rather because its opponents eventually die, and a new generation grows up that is familiar with it."

Be that as it may, the Earth is not a simple cooling body, and the radioactive isotopes of uranium, thorium, and potassium in its outer crustal rocks do provide a substantial portion of the heat that flows out of it. But even so, thanks to a lack of knowledge about the exact amount and distribution of such radioactive elements in the Earth, calculations for the age of the Earth based on the Earth's heat flux still lack precision.

In due course, radioactive decay would provide the definitive method for finding the Earth's age as well as for determining the various stages within the geologic record. But even without the precision with which physicists could reckon such things once they had the tool of radioactivity, geologists had done quite well.

From its earliest days geology knew of the three distinct eras, each with its own distinct faunal assemblage, and each separated from the next by a massive faunal turnover (or mass extinction). The accompanying figure, from 1860, shows the three eras of the Phanerozoic (meaning "evident life"). Life in the Paleozoic ("ancient life") consisted mainly of aquatic forms; in the Mesozoic ("middle life"), reptiles predominated; and the Cenozoic ("modern life"), is known as the age of mammals. As we have seen, two mass extinction events (and they are the two greatest mass extinction events in the geologic record) separate these three eras: the Permian-Triassic (or P-Tr) boundary and the Cretaceous-Tertiary (or K-T) boundary.

As these three recognizable eras were further divided into periods, epochs, and ages, each subdivision has its own distinct faunal assemblage, and some, though by no means all, are separated by extinction boundaries. These various divisions, particularly those around Cretaceous-Tertiary time, are delineated in the two figures on pages 17 and 18.

How did physicists and geologists collaborate to assign *dates* for these divisions? We must start by reviewing basic considerations

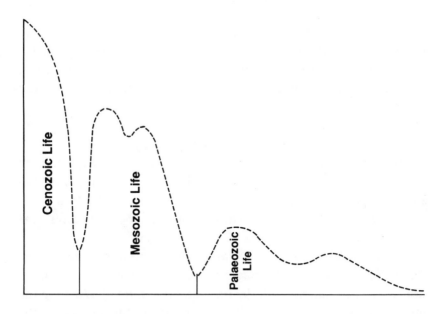

The relative number of life-forms in three eras, as understood in the nineteenth century. *Source:* **Phillips 1860.**

from atomic physics. Any physical element is defined by the number of protons in its nucleus, which, in a neutral atom, equals the number of its associated electrons and is known as its atomic number. Thus, argon is element 18, with 18 protons in its nucleus. Each element can also have varying numbers of other elementary particles called neutrons in its nucleus, and these variations are called isotopes. Adding up the protons and neutrons in a given isotope gives its mass number. Thus, argon-40 is an isotope of element 18: it has 18 protons and 22 neutrons in its nucleus.

Some isotopes are stable, but some are *unstable,* meaning that they decay by emitting elementary particles and become other (stable) isotopes. The rate of decay is proportional to the number of atoms present at any time. A convenient measure of this decay process is the *half-life,* defined as the amount of time it takes for an initial quantity of the unstable element to be reduced to half that amount. The same amount of time is required for the half to be reduced to one-fourth of the original amount, and so on in an exponential manner.

This rate of decay—an element's half-life—is what is used to determine the age of a rock or, more elaborately, the Earth itself. For example, take an igneous rock that has cooled down to its solid state from an initial melt. Say that a geochemist finds that it contains a certain amount of argon-40. He knows that potassium-40, an unstable isotope, decays to form the stable isotope argon-40. And he knows, thanks to the physicists, that the half-life of potassium-40 is a little more than a billion years. So he can add up the amount of the two isotopes and determine how much potassium-40 was present in the rock when it first cooled into a solid. Then, with potassium-40's known half-life, he can determine when that decay process first started.

Of course, it's not quite that simple. Argon-40 comes in the form of a gas, and some may have been present in the original molten material along with the original potassium-40. Also, in the interim, some of the argon-40 gas could have diffused out of the rock. But by trial and error, geochemists have found that certain minerals like hornblende and muscovite are more retentive than others, and so they use them for such analyses. Even so, later thermal events could have driven off some or all of the argon-40 produced up to that time and reset, so to speak, the potassium-argon clock. So potassium-argon age determinations are likely to err on the side of a younger age than is the actual case. Nevertheless, using this "clock" and many others involving other elements, scientists have delineated the actual ages and time intervals in the geologic record—within certain ranges of error, to be sure, but the error tends to be reduced as techniques become more refined.

The oldest known rocks solidified about 3.8 billion years ago; the Phanerozoic eon began 570 million years ago. The range of error in such ancient times can be from five to ten million years. As we come closer to the present, the range of error decreases. In 1983, for example, the K-T transition was assigned an age of 66.4 million years, but today it is taken to be closer to 65 million years.

All this radiometric dating is fine, so long as one has igneous rocks to date. But many sedimentary sections—which are of prime interest to the paleontologist—have few if any such volcanic layers associated with them. Happily, there is another timescale they can use. As noted previously, the Earth acts like a giant magnet, in essence a huge electrodynamic dynamo, resulting from the fluid motion of molten conducting material within its core. The fluid motion is controlled in part by the Earth's rotation, and the magnetic pole is roughly the same as the geographic pole. As one might expect from so dynamic a system, the fluid motion has varied over time. The magnetic north pole has wandered slightly over time; its strength has decreased by about five percent in the past century.

Throughout geologic time, the variations in the magnetic system have been even greater. From time to time the magnetic north pole has undergone a great reversal, becoming the south pole and then after some interval becoming the north pole again. *This* sequence can be dated by radiometric means, providing a secondary geologic time standard, and it is this timescale that has proved most fruitful in dating the events recorded in sediments deep in the ocean. There among the sinking detritus of tiny life-forms, particles containing a few grains of magnetite and other magnetic materials have floated down in a relatively quiet environment and lined up on the ocean floor according to the Earth's magnetic field. Once they have been covered and cemented to neighboring particles, they are preserved like miniature compasses, reliably indicating the direction of the Earth's magnetic field at the time they settled there. When a core is taken through such sedimentary rock, the sequence of magnetic reversals can be determined and then compared with the standard reversal sequence, thereby providing an accurate date for each level within the core.

The K-T transition occurred during an interval of a reversed magnetic polarity—that is, when the present geographic north pole was a magnetic *south* pole. Within the numerical sequence of reversals that have been determined over geologic time, it is labeled 29R (reversed), which was preceded by 29N (normal) and followed by 30N. Magnetic interval 29R has been found to have lasted for 470,000 to 610,000 years. The preceding interval, 29N, lasted from 560,000 to 790,000 years. And 30N lasted from 1,390,000 to 1,820,000 years. This dating is crude, to be sure, but nonetheless a great deal more rigorous than counting petals on a daisy. An important point to be remembered is that by the time the Alvarezes announced their hypothesis, a great deal of rigorous science concerning the great marine and other extinction events that took place around the time of the K-T transition already existed.

It is time now to take a close look at what Dartmouth geologist Charles Drake has called *paleonecrology,* the extinction record itself. We want to draw attention to several important distinctions that geologists deal with, both having to do with time. In a sense, for a geologist, time slows down. Some extinctions (of an entire order, for example) take place over several million years, and geologists think of this extinction rate as gradual. Others may take place over a period of a few hundred thousand years, and while a nongeologist would be forgiven for thinking that a few hundred thousand years is an unimaginably long interval, a geologist would call an extinction occurring in such a brief time-frame catastrophic.

Second, there is an important conceptual difference between *relative time* and *absolute age* (which radiometric dating tries to establish). We sometimes can tell if a group of animals became extinct before (or after) another group without knowing the exact time when either group disappeared. Many of the definitive extinction ages that we have for ancient times come from sedimentary rock sections of oceanic origin—sections that have subsequently been lifted up and exposed on land, or sections drilled from present deep ocean basins. (These latter sections have been retrieved with amazing thoroughness by the Deep Sea Drilling Project, originally a U.S. project but now an international program, which began in 1964.)

In both cases, most of the sediments are simply the result of calcareous plankton, whose remains drift down to the ocean bottom. The remains accumulate at varying rates, but a rate of one centimeter per one thousand years is typical. So a one-thousand-year interval is about as precise as we can get by way of time discrimination in such sections. On the other hand, if chemical conditions in the ocean are (from the scientist's standpoint) adverse, all the calcium carbonate remains will dissolve; this leaves no fossil record at all, but just fine clay particles that have drifted down to the bottom. The sedimentation rate for such particles is about a tenth of a centimeter per thousand years or less.

These neat layers can be severely distorted. Near shore, land runoff can make the sections substantially thicker, and periods of erosion can eliminate whole layers. Another plague for the paleontologist is what is called the Signor-Lipps effect, named for Philip Signor and Jere Lipps of the University of California at Davis, who first described it. In a given geologic section, there may simply be a scarcity of fossil remains; no matter how carefully one samples such a section, the species that were less prevalent in life will normally disappear in the section before those species for which there are more abundant remains. The uppermost or "highest occurrence" of a given species in the stratigraphic record cannot always be associated with its "last appearance on Earth."

In any event, for a variety of reasons, hiatuses do appear in the record. If a hiatus occurs during a period when a given species was gradually going extinct, what we will see is an abrupt, maybe catastrophic change. In such a case, we have no way to tell how the extinction took place. For example, the geologic section at Gubbio, where the Alvarez team found the iridium anomaly, consists of limestone beds (carbonate plankton remains averaging about 20 to 30 centimeters in thickness) alternating with barren clay layers about one to two centimeters thick. Each limestone layer took, on average, between 20,000 and 30,000 years to accumulate, each clay layer between 10,000 and 20,000 years. In all, this limestone-clay sequence extends over a period of some 50 million years, from the Cretaceous well into the Tertiary. One of the clay layers—the one with the iridium anomaly—is preceded by a limestone layer with characteristic Cretaceous species, while the succeeding limestone layer has characteristic Tertiary species. The barren clay layer in between corresponds to a K-T transition, certainly. Even so, we are left interpreting a period of 10,000 to 20,000 years in the clay layer. In itself, then, this clay layer tells us very little about how the extinctions took place.

In other words, from the standpoint of foraminifera species, what *exactly* happened during that revolutionary period is

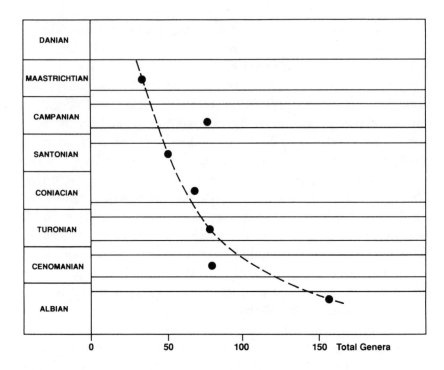

The continuous decline of ammonites during the Cretaceous period. *Source:* **Wiedmann 1969.**

unknown from this section. Paleontologists could get a better idea if they could find a layer of limestone somewhere else that corresponded precisely with the layer of clay that, at Gubbio and elsewhere, occurred at exactly the time the Cretaceous era was succeeded by the Tertiary—that is, at the K-T boundary.

Even with these complexities, paleontologists in 1980 still had a good idea of the relative as well as absolute timing involved in the extinctions of marine creatures in K-T times. The ammonites showed a pronounced decline in species diversity beginning in Albian time, some 100 million years ago, and continuing up to the K-T boundary, 65 million years ago. Some of this decline had, by 1980, been attributed to a succession of sea-level regressions that would have eliminated the ammonites' shallow-water habitats here and there. In a geologic section in southwestern France, the ammonites showed a progressive decline in species over the

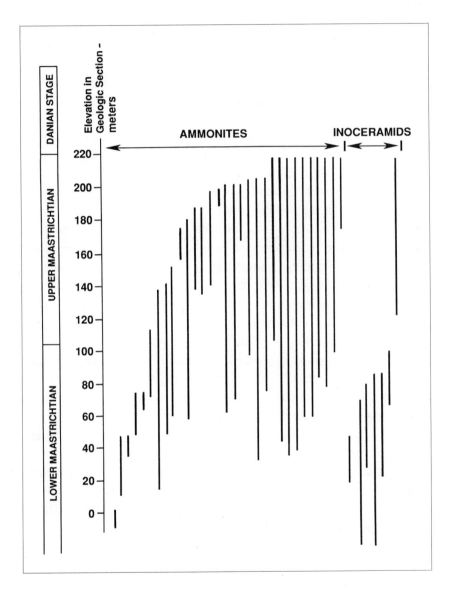

The stratigraphic extent and ages of various species of ammonites and inoceramid bivalves for late Cretaceous Bay of Biscay geologic sections, southwestern France. *Source:* Ward et al. 1991.

last eight million years of the Cretaceous period (that is, in Maas-trichtian time), and the final species vanished at or within 10,000 years of the microfossil extinctions that define the K-T boundary. Similarly, the remains of *inoceramids* (counterparts of bivalves of today) show a comparable decline over the last 25 million years of the late Cretaceous, with all but one enigmatic species disappear-ing four million years before the boundary.

The K-T boundary is usually defined on the basis of the micro-fossil (that is, foraminiferal) extinction record, and paleontologists had found that even here the fossil record does not represent an instantaneous event. Instead, sequential extinctions (not one but several) occurred stepwise over some 400,000 years. All this and a great deal more was known about the last years of the Cretaceous by geologists and paleontologists when, in 1980, the Alvarez group unleashed its hypothesis; so again, it is no wonder that most of them thought Alvarez was providing a wrong answer to a wrong question. There was no paleontological evidence for an instanta-neous extinction event for those species—in all, 50 percent of *all* extant species—around K-T times. Not for dinosaurs, nor flying reptiles. Not for marine reptiles, nor shallow-water shellfish, nor microfossils, nor nannofossils. There was no big dinosaur bone pile, no great midden of shellfish remains that might have resulted from an instantaneous extinction event. Not only that, such a "lights out" scenario hardly accounted for the K-T survivors—crocodiles, turtles, snakes, lizards, insects, birds, and mammals, not to mention some fish and other marine forms. The whole thing seemed absurd, but given the instantaneous and wide publicity the Alvarez hypoth-esis received, it had to be contended with. Dissent—however little it reached the public—was quick to occur.

FIVE
Early Dissent

EXPRESSIONS OF DISSENT TO THE ALVAREZ hypothesis were immediate. Some were based on the obvious gradualism of the extinctions in question, and further studies would emphasize this aspect of the extinction record. Other objections were based upon a wide body of knowledge that was already accumulated about a variety of environmental and geologic events that took place around the time of the K-T transition—events that of themselves could explain the mass extinctions. Yet other objections concerned Alvarez's killing mechanism—a gigantic dust cloud thrown up by the impact of a gigantic meteorite. Alvarez estimated that the dust cloud was a thousand times greater than the one that followed the eruption of Krakatoa in 1883. This eruption, too, was something scientists already knew plenty about.

The uninhabited island of Krakatoa, located between Java and Sumatra, was a chief navigational aid to sailors plying the Sunda Strait when suddenly, in 1883, it was lost to view. In August, on the 26th and 27th, eruptions nearly obliterated the island and set in motion a train of damage. More than 30,000 people were swept off Java and Sumatra by a sea wave called a *tsunami,* a low wall of water caused by the sudden vertical movement of the seafloor, often the result of earthquakes near deep sea trenches. These waves can propagate across the ocean, in the physicist's cool parlance, as "low-amplitude and long-wavelength disturbances," but as they approach the shallow waters near land, their amplitude increases. In other words, as a tsunami nears land, it rises up as a rushing wall of water some ten or more feet high. The one that followed after Krakatoa blew was caused either by ejected material plummeting into the sea or by the collapse of the island structure itself.

The Royal Society of London was able to catalogue more distant results as well, making this eruption the first geophysical event to be studied scientifically on a global scale. To begin with, it made one of the loudest sounds in history, being heard as far as 3,000 miles away on the island of Rodriguez in the Indian Ocean, where chief of police James Wallis reported that "several times during the night of the 26th–27th reports were heard coming from eastward, like the distant roar of heavy guns." Low-frequency sound waves circled the Earth as many as seven times, and were picked up by barometric pressure gauges at stations around the globe.

For a year after the eruption, many ships' logs noted floating pumice in the Indian Ocean. In March 1884 Captain Gray of the *Parthenope* found the central Indian Ocean strewn with pumice that was covered with barnacles, testimony to its long residence in the water.

More lasting were the atmospheric effects of dust and sulfate aerosols that had been injected into the stratosphere. For three years in many parts of the world, the days were filled with a blue or green haze, with spectacular red glows just after sunset and just before dawn. In Poughkeepsie, New York, there was such "an intense glow in the sky that fire engines were called in the morning" on November 27, 1884—more than a year after the eruption.

An even greater eruption—the greatest in recent historical times—was that of Tambora on Sumbawa Island, adjacent to

Java, in the Indonesian archipelago in 1815. At least ten times greater than Krakatoa, it was described definitively by Sir Thomas Stamford Raffles, founder of Singapore and British Resident in Malaya and the East Indies at the time of the eruption.

It began on the 5th day of April, and was most violent on the 11th and 12th, and did not entirely cease until July. The sound of the explosion was heard in Sumatra, at a distance of nine hundred and seventy geographic miles in a direct line, and at Ternate, in the opposite direction, at the distance of seven hundred and twenty miles.

Out of a population of twelve thousand, only twenty-six individuals survived on the island. Violent whirlwinds carried up men, horses, cattle and whatever else came within their influence into the air, tore up the largest trees by the roots, and covered the whole sea with floating timber. . . . On the side of Java, the ashes were carried to a distance of three hundred miles, and two hundred and seventeen toward Celebes, in sufficient quantity to darken the air. The floating cinders to the westward of Sumatra formed, on the 12th of April, a mass two feet thick and several miles in extent, through which ships with difficulty forced their way.

The darkness occasioned in daytime by the ashes in Java was so profound, that nothing equal to it was ever witnessed in the darkest night. . . . Along the sea-coast of Sumbawa, and the adjacent isles, the sea rose suddenly to the height of from two to twelve feet, a great wave rushing up the estuaries, and then suddenly subsiding. Although the wind at Bima was still during the whole time, the sea rolled in upon the shore, and filled the lower parts of houses with water a foot deep.

The area over which tremulous noises and other volcanic effects extended was one thousand English miles in circumference, including the whole of the Molucca Islands, Java, a considerable part of Celebes, Sumatra and Borneo.

Dust and sulfate aerosols injected into the stratosphere by Tambora spread out globally, shielding incoming sunlight. In New England the following year, 1816, was called "The Year Without a Summer" and also, with Yankee wryness, "Eighteen Hundred and Froze to Death." The average temperature in June was seven degrees Fahrenheit below normal, crop failures were rife, and prices soared. Things were even more disastrous in Europe. Crop failures were universal, and food shortages led to local famines and anarchy.

Even larger eruptions occurred in earlier times. In Sumatra some 75,000 years ago, the eruption of the Toba volcano left a collapse structure (or *caldera*) 30 by 60 miles in dimension. Ash was spewed over much of the Indian Ocean, leaving a layer on the bottom as much as four inches thick 1,300 miles away. Six hundred thousand years ago, in what is now Yellowstone National Park, an eruption spread ashes a few inches thick over most of the present United States west of the Mississippi. Both of these eruptions may well have wiped out most of their regions' flora and fauna as well as causing havoc on a global scale, but both regions recovered in a relatively short period.

In a brief report in the February 13, 1981, issue of *Science,* Dennis Kent of Columbia University estimated that the Toba eruption was four hundred times as powerful as Krakatoa but that no massive extinctions or other extraordinary effects on life are associated with it. Alvarez was quick to respond, writing that "Kent estimates that the Toba eruption would have ejected about 400 times as much material as Krakatoa did, close to our estimate of 1000 times Krakatoa for the impact event, although Toba did not produce extinctions. This consideration may make it possible to place a lower limit on the size of extinction producing events."

From a physicist's standpoint this lower limit might make sense, but it is unlikely that a biologist would accept the existence of an abrupt cutoff point somewhere between an event four hundred times Krakatoa and a thousand times Krakatoa, with the first producing no noticeable biotic effects and the second producing the extinction of 50 percent of all species on a global scale.

So much, at least for the time being, for the weapon—the impact and its aftermath. In 1980 it seemed to many to be of insufficient power to accomplish the result attributed to it.

Paleontologists, meanwhile, took up the Alvarez argument. They were satisfied that the immediate cause of the demise of dinosaurs in western North America had been known for some time. The climate and the dinosaurs' habitat had changed with the disappearance, during the last part of the Cretaceous, of the interior seaway that had extended from the Gulf of Mexico to the Arctic Ocean. Robert Sloane of the University of Minnesota and others pointed out that the dinosaur extinction in Montana, Alberta, and Wyoming was a

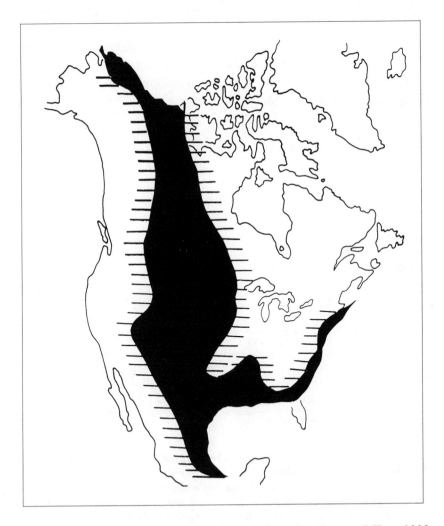

Late Cretaceous seaway in western North America. *Source:* **Officer 1990.**

gradual process that began seven million years before the end of the Cretaceous and accelerated rapidly in its final 300,000 years. They attributed the gradual decline to changes in sea level and climate, and the more accelerated decline at the end to competition for food with the new and expanding population of plant-eating mammals.

In another study, Michael Williams of the Cleveland Museum of Natural History reviewed fossil data from the prolific dinosaur fields of Montana for the final few hundred thousand years of the Cretaceous. As shown in the diagram, there is a regular and systemic decrease in the diversity of dinosaurs, as represented by the types of teeth and by their absolute number extracted from the channels of old river beds. Generally speaking, there is a greater chance for the preservation of teeth than bones in these channels. Williams concluded his review by writing

> The distribution of the remains in the rock record is compatible with a gradual extinction, but could hardly be

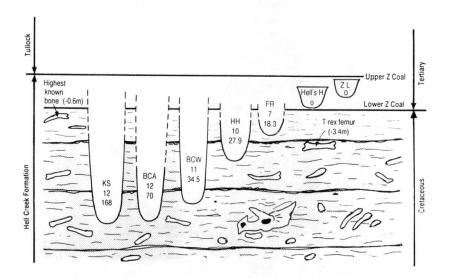

Dinosaur bones and teeth in Montana. In the Hell Creek formation, bones of dinosaurs become progressively rarer toward the K-T boundary, which is represented by "Lower Z Coal" (but no iridium anomaly). The number of dinosaur teeth declines in a sequence of seven paleochannels at Hell Creek; for example, channel KS had 168 dinosaur teeth per ton, representing twelve separate tooth fauna. *Source:* Williams 1994.

further from the predictions of a catastrophic extinction. Not only is there no increase in the number of remains encountered as we approach the iridium layer, but the remains become rarer until not a single unequivocal occurrence is known at the time the iridium was deposited.

In this same area of the world, an abrupt transition from fernlike plants to flowering plants occurred at about the same time as the observed K-T iridium anomaly, but the floral change in the region didn't occur at the same time as the end of the dinosaurs. In Alberta, John Lerbeckmo of the University of Alberta and others found that the last appearance of *Triceratops,* which was the last surviving plant-eating dinosaur, occurred a meter or so *below* the evidence of the floral change, and the last appearance of the last carnivorous dinosaur, *Tyrannosaurus,* occurred somewhat lower than *Triceratops.*

Again in Montana, similar results were documented by David Archibald of San Diego State University and William Clemens of the University of California at Berkeley. If, they emphasized, the iridium anomaly in this part of the world was in fact caused by an impact, it was difficult to associate the impact with the end of the dinosaurs because that end had already occurred. The extinction of the western North American dinosaurs may well have been a regional event—again, probably the result of the loss of their inland sea—and possibly an event without global implications. Elsewhere, different local or regional conditions might have been the cause.

In any event, western North America and northern Europe appear to have been the last refuges for dinosaurs. Elsewhere in the world, their last appearance in the fossil record was some seven to fifteen million years before the K-T boundary. The last known dinosaurs in South America were Campanian to middle Maastrichtian in age, as were the Asian dinosaurs. In southern Europe the last ones appeared in the Maastrichtian as well, and in northern Europe only one and possibly two species survived into the Maastrichtian. (There is a caveat here: Except for western North America, there is a paucity of dinosaur fossil data for the late Cretaceous, and the seven-to-fifteen-million-year hiatus else-where may be the result of a lack of defining information.)

It would appear that not only were dinosaurs well on the way to oblivion before the end of the Cretaceous, but their decline was nothing in terms of rate or magnitude compared to the many extinction events that had previously befallen dinosaurs. In early periods in the Mesozoic, dinosaur turnover rates were high and occurred over the course of a few million years or less. What does stand out during the late Cretaceous is a fall in the *origination* of new species to compensate for the extinction rate of older species. Perhaps the scholars of Michael Benson's nonquestion phase were right after all: the dinosaurs died from some kind of racial senility, simply running out of the genetic variability necessary to evolve and survive as environmental conditions changed.

Beyond all that, the K-T boundary itself is geologically defined on the record of marine invertebrate fauna and *not* on dinosaurs. Here again extinctions are known to have been gradual rather than instantaneous. For example, William Zinsmeister of

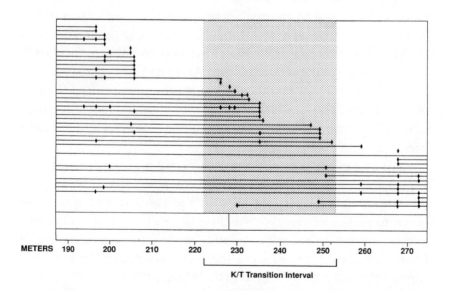

At Seymour Island, Antartica, extinctions of ammonites, bivalves, and other sea creatures occurred gradually—before, during, and after the K-T transition interval. *Source:* Zinsmeister et al. 1989.

Purdue University and his colleagues studied the record of macro-fossil extinctions (ammonites, bivalves, gastropods, lobsters, and the like) on Seymour Island, Antarctica. As shown in the accompanying diagram, they found a transition interval of some 30 meters over which these extinctions occurred, representing a period estimated to have lasted from a half-million to a million years.

The K-T boundary, as we have noted earlier, is usually defined on the basis of the microfossil record—that is, the record of extinctions among foraminifera. Even here, paleontologists found nothing to suggest an instantaneous global event but instead sequential extinctions that occurred stepwise over 400,000 years. Gerta Keller of Princeton University was one of these, and not long after the Alvarez hypothesis became doctrine in the minds of some, she published data to the contrary based on microfossil transitions at the Brazos River in Texas and at El Kef in Tunisia.

At the Brazos River, nearly half of the Cretaceous species (46 percent) disappeared in a first extinction event, and another third disappeared in a second event. What were left of the species (some

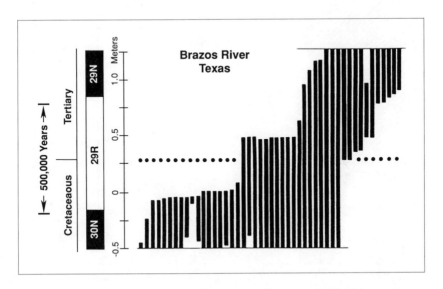

Foraminiferal extinction record across the K-T transition at Brazos River, Texas. Magnetic polarity interval 29R was approximately 500,000 years in duration. Several iridium anomalies appear in the Brazos River section; the level of one of these is shown by the dotted line. *Source:* **Keller 1989.**

21 percent) survived the K-T boundary. At El Kef, thirteen species of foraminifera (or 29 percent of the Cretaceous species) disappeared between twenty-five and seven centimeters below the level of an iridium anomaly, with twelve species (or 26 percent) disappearing at the iridium anomaly level. Five species (or 11 percent) disappear within the fifteen centimeters above the iridium level, and eight (or 17 percent) disappeared at higher levels. Only 17 percent remained—totaling eight species—as K-T survivors.

There have also been several studies of the K-T nannofossil (calcareous phytoplankton) extinction record. One such detailed study was conducted by Hans Thierstein of the Scripps Institution of

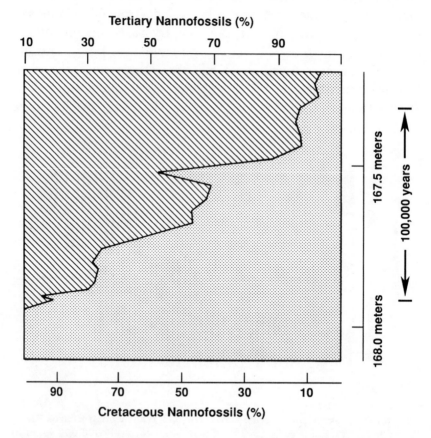

Gradual shift in abundance of Cretaceous and Tertiary nannofossils versus core depth at DSDP site 384. *Source:* **Thierstein and Okada 1979.**

Oceanography and Hisatake Okada of Yamagata University, on a core section from Deep Sea Drilling Project site 384 in the western North Atlantic. The change from Cretaceous nannofossils to Tertiary nannofossils occurs gradually over 60 centimeters of the core. The estimated sedimentation for this core interval is 0.6 centimeters per thousand years, so that the nannofossil change represents a time span of 100,000 years—somewhat less than is usually associated with the microfossil transition but still hardly instantaneous.

It seemed clear enough in 1980 and still clearer as further studies emerged, such as Gerta Keller's, that even if a meteoritic impact occurred at K-T time, it simply could not explain the extinction record. Half the world's creatures had simply not folded up and slipped away on July 16, 65 million years ago.

Not only did the record show mass extinctions before the K-T boundary, but there were perfectly good reasons to explain, at least in rough detail, the mass extinctions themselves. Certainly they were not the utter mystery that the Alvarez hypothesis, however implicitly, suggested was in need of a deus ex machina in the form of a giant meteorite from outer space to clear up.

No one would deny that the extinctions that occurred around K-T time were extraordinary (though importantly, they were not unprecedented). Some 50 percent of all species globally disappeared, a proportion that is yet to be equaled in the many millions of years since. Extraordinary events of a global nature must have been the cause. In fact, several such events had been evident in the geologic record for some time.

One such event was a major drop in sea level during the late Cretaceous, beginning about 70 million years ago. This drop was followed by a complementary rise shortly after the K-T boundary, during which the Earth "recovered" to its normal condition. The second extraordinary event was an intense period of global volcanism that occurred over a shorter interval before, during, and after the K-T boundary.

Geologists track a fall in sea level on a stratigraphic core section by following the sequence of sediments from marine to shallow water to terrestrial. Sediments that occur in the opposite sequence denote a rise in sea level. Given the nature of such cores, it is difficult to ascertain how much the sea level changed. One guess made from a section in Venezuela was 50 to 100 meters. Another estimate came from a section from the Arabian Peninsula, and as the figure below shows, it suggested a gradual rise beginning 90 million years ago, followed by the rapid fall during Maastrichtian times.

Once again, geochemistry has come to the rescue—in the form of a strontium-isotope anomaly in the geologic record in this period. *This anomaly is as striking in its own way as the iridium*

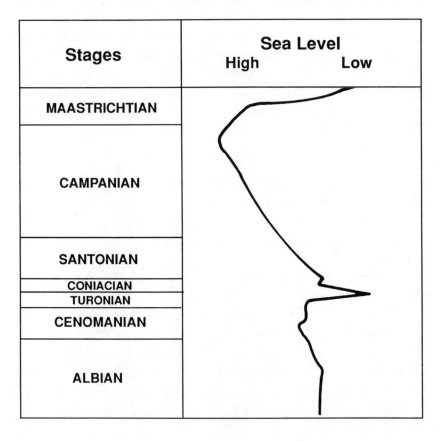

Sea-level changes during the Cretaceous on the Arabian Peninsula.
Source: **Hallam 1992.**

anomaly. In brief, it is an anomalous ratio of strontium-87 to stron-
tium-86 that occurred from 69 to 65 million years ago.

Strontium in seawater is taken up by calcareous plankton in
building their shells, and from these shells, strontium levels in the
ocean can be determined. The ratio of strontium-87 to strontium-86
is higher in continental runoff but lower in the hot waters associated
with deep ocean ridges where volcanic activity takes place. The ratio
for oceanic waters, in general, is a mix of these two ratios. So an
upward blip in the ratio in the oceans of the late Cretaceous must
have been caused by increased runoff from the continents—which in
turn was the result of increased erosion associated with the fall in sea
level at that time. If one goes through all the strontium-isotope calculations, one finds, as in the following figure, that the regression
phase occurred from 69 to 65 million years ago, a date that happily
agrees with the estimates based on sedimentary sections.

At the same time that the sea level was falling, the Earth
was experiencing intense volcanism, which is evident today most

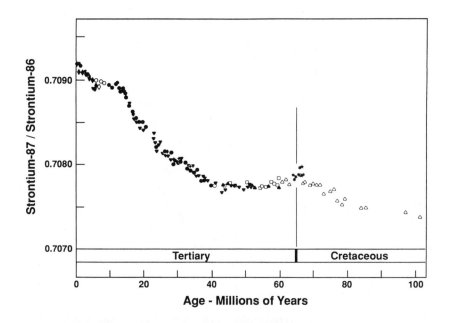

**Measured strontium-87 to strontium-86 isotope ratio, plotted against
stratigraphic age. *Source:* Hess et al. 1986.**

prominently in western India, in the continental flood basalt deposits known as the Deccan Traps. Radiometric and paleomagnetic analyses have shown that the flood basalts were deposited over a relatively short period of time around the K-T boundary. The eruptions began during the magnetic polarity interval 30N and continued during 29N, but the bulk of the basalts were extruded during 29R.

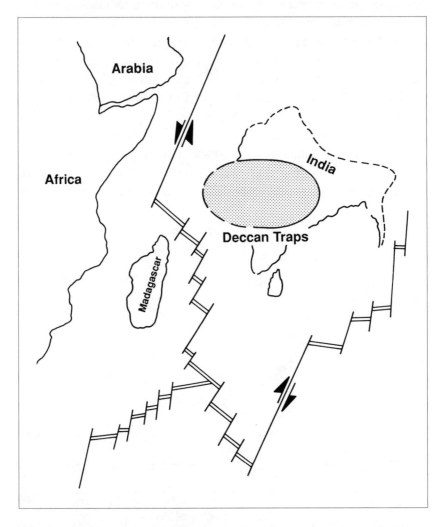

At K-T time, during the eruption of the Deccan Traps, India was part of a tectonic plate moving northward off the present coast of Africa. *Source:* Adapted from Courtillot et al. 1986.

And it was during the 500,000-year-period of 29R that the transition from Cretaceous to Tertiary microfossils occurred. That is to say, the major K-T mass extinctions occurred then.

The Deccan volcanism presumably began as a violent sequence of explosive events and continued as a more quiescent outpouring of lava. In all, more than a million cubic kilometers of lava was ejected there. Nor was the intense volcanism confined to what is now India—it was evidently global in nature. In North America intense volcanism is associated with the mountain-building epoch when the Rocky Mountains came into existence and the inland sea disappeared, a disappearance that was hastened during Maastrichtian times, as the global sea level dropped.

To most geologists, such titanic changes in the Earth's overall environment seemed ample disruption to have caused mass extinctions. The volcanism would have ejected enormous amounts of volatiles such as sulfur dioxide, carbon dioxide, and chlorine into the atmosphere and stratosphere, as well as spreading them about the terrestrial environment. Before 1980, Dewey McLean of Virginia Polytechnic Institute had suggested that a terminal Mesozoic greenhouse effect had resulted from the Deccan Traps' carbon dioxide emissions. Through the years he has been a strong advocate of volcanism as a powerful agent in the K-T extinctions. Another scientist, McKenzie Keith of Pennsylvania State University, had suggested that the chlorine emissions would have destroyed the protective ozone layer, resulting in an increase in damaging ultraviolet radiation on Earth. The bare-skinned dinosaurs, he pointed out, would have been at greater risk in such a situation than birds and mammals, which were protected by feathers and fur—an insight that might explain the selectivity of the terrestrial extinctions.

Nothing like these events, or the extinctions many scientists thought they explained well enough, have occurred since K-T times. But much the same events seem to have occurred during the time of an earlier mass extinction, when the Permian period merged into the Triassic some 250 million years ago. Indeed, this transition, the P-Tr boundary, marks the end of the Paleozoic period and the beginning of the Mesozoic. There seems to be a pattern in these events, spread apart by such a long period of time; we will return to them in the final chapter of this book.

Vincent Courtillot and colleagues at the University of Paris and, in a separate investigation, Robert Duncan and Douglas Pyle of Oregon State University, carried out the precise radiometric and paleomagnetic dating of the Deccan Traps that demonstrated their coincidence with the K-T boundary. Both groups have emphasized the potential importance of these volcanic erruptions toward understanding the K-T extinctions. Anthony Hallam of the University of Birmingham in England described the rapid transgression-regression cycle of the late Cretaceous to early Tertiary and emphasized its importance toward understanding the extinctions that immediately preceded the K-T boundary.

Both before the Alvarez impact hypothesis and afterward, suggestions that these major global events explained the mass extinctions at the K-T boundary had been made, but none had received much public attention. The Alvarez camp variously ignored them, along with the evidence that the mass extinctions had been relatively gradual events, or attacked the competence and personalities of their advocates. It was at best a rancorous controversy, and rife with an odd kind of politics, as we shall explore more fully in the next chapter. With all the rancor, there was little room for compromise.

At a meeting of the Society of Vertebrate Paleontologists in 1985, for example, William Clemens commented on some finds near Prudhoe Bay in northern Alaska, by way of refuting that a curtailment of photosynthesis by the meteorite impact led to the dinosaur die-out. Dinosaurs were already accustomed to prolonged darkness, he said, from the long Arctic night.

> The challenge in this is to discover how dinosaurs adapted to the rigorous seasonal regime of daylight and darkness in the polar regions. I don't believe they survived by migrating twice a year, because the distances would have been too great for them. With the photosynthetic growth of plants cut off by

darkness, herbivorous dinosaurs may have just shut down for the winter, or found some other way to reduce their food requirements. But survive they did, as we see in the fossil record. What does that say about the impact theory of extinction? It's codswallop.

In return, Luis Alvarez simply sniffed that Clemens, who is a paleontologist, was not qualified to make statements about the fossil record.

An informal survey by *The New York Times*'s Malcolm Browne at the meeting showed that only four percent of the paleontologists there thought an extraterrestrial object had caused the demise of the dinosaurs, though nine out of ten were willing to accept the possibility that an asteroid might have hit the Earth around that time. A noted dinosaur specialist at the meeting, Robert Bakker of the University of Colorado Museum, had this to say:

The arrogance of these people is simply unbelievable. They know next to nothing about how real animals evolve, live and become extinct. But despite their ignorance, the

William Clemens. *Source:* **William Clemens.**

geochemists feel that all you have to do is crank up some fancy machine and you've revolutionized science. The real reasons for the dinosaur extinctions have to do with temperature and sea-level changes, the spread of diseases by migration and other complex events. In effect, they're saying this: "we high-tech people have all the answers, and you paleontologists are just primitive rockhounds."

Indeed, Malcolm Browne quoted Luis Alvarez as saying, "I don't like to say bad things about paleontologists, but they're really not very good scientists. They're more like stamp collectors."

In an article in *American Scientist* in 1988, Keith Stewart Thomson, president of the Academy of Natural Sciences and former professor of biology and dean of the graduate school at Yale University, gave a useful list of the procedures by which a radical idea becomes a new orthodoxy (or not, as the case may be). First there must be a *problem without a solution.* Then comes a *breakthrough,* the dream of every scientist—in this case it was the iridium anomaly. The next step is *spreading the word*—the association of the iridium anomaly with dinosaurs quickly saw to that. The next step is the *excommunication of the apostate,* followed by *reaction* from the entrenched community, followed by the *mobilization of resources* to demonstrate the appropriateness of the new idea. Then, as the game gets more complex, comes *the changing of the terms of the discussion,* often followed by a *silly season,* which includes the proposal of new ideas even more far out than the original one, at the end of which *calmer heads prevail.* So far in this discussion, we have seen a problem raised that Alvarez and his colleagues decided was without a solution, We have seen their breakthrough—iridium!— and the spreading of the word. And we have seen some of the next step—excommunicating the apostate—in the contempt with which early dissent, if it was heeded at all, was treated.

As we shall see in the following chapters, the rest of Thomson's formulation would unfold in a grand if unseemly mix of politics and science.

SIX
Science and Politics

ONE DAY A HISTORIAN OR SOCIOLOGIST OF SCIENCE may collect all the anecdotes, personal letters, interoffice memoranda, and other nonpublic documents that surrounded this controversy and, from them, draw an embarrassing and unseemly portrait of scientists at work and at each other's throats. Based on such documents, it may even be possible to put together a picture of what motivates some scientists to such rancor. That is not the purpose of this book, nor does such an investigation lie within the talents of its authors, whose interest lies more firmly in the scientific merits of the case, broadly speaking.

Nevertheless, from time to time, public reports of the ad hominem aspects of the controversy have appeared, and one of these, a report in *The New York Times* in January 1988, is worth reviewing, if only to show that an active excommunication effort did take place. In its report, the *Times* reported an interview with Luis Alvarez, at a time when he had been diagnosed with terminal throat cancer. Alvarez spoke openly and specifically about two of the earliest and foremost critics of his impact hypothesis.

One of these was Dewey McLean who, as early as 1978, had suggested in *Science* that high levels of carbon dioxide in the atmosphere at the end of the Cretaceous period had caused a global greenhouse effect; the higher temperatures could have brought about the final dinosaur extinction by interfering with their reproductive processes. Subsequently, McLean theorized that the high levels of carbon dioxide came from the gigantic volcanic upheaval that produced the Deccan Traps. He went on to speculate that such volcanism might account for the iridium anomaly that had triggered the Alvarez hypothesis. As a result of his work, McLean charged, Alvarez's allies had made an attempt to undermine his career at Virginia Polytechnic Institute.

"Several scientists," the *Times* article said, "who requested anonymity said in interviews that scientists in the Alvarez camp subsequently tried to intercede with officials at Virginia Polytechnic Institute to block his promotion to full professorship and to discount his work. Dr. McLean did receive the promotion."

When asked about this incident, Alvarez denied the charges, but according to the *Times,* he added the following comment:

> If the president of the college had asked me what I thought
> of Dewey McLean, I'd say he's a weak sister. I thought
> he'd been knocked out of the ball game and had disappeared,
> because nobody invites him to conferences anymore.

In the same interview, Alvarez discussed his colleague at Berkeley, William Clemens, the paleontologist who had found dinosaur remains on Alaska's North Slope and concluded that dinosaurs were somehow capable of surviving long periods of Arctic night. Clemens had thereby rendered the notion of a long darkness caused by an impact inapplicable, and he had called the Alvarez hypothesis "codswallop." According to the *Times,* Alvarez said "that he considers Dr. Clemens inept at interpreting sedimentary rock strata and that his criticisms can be dismissed on the grounds of general incompetence."

This seems harsh, hardly the kind of discourse the public has come to expect from the collegial realms of science, but as the *Times* quoted Alvarez: "I can say these things about some of our opponents because this is my last hurrah, and I have to tell the truth. I don't want to hold these guys up to too much scorn. But they deserve some scorn, because they're publishing scientific nonsense."

There is more to his remarks, however, than even Alvarez's unfortunate medical condition. He evidently had a rather difficult side, tending to be brutal with those who disagreed with him and intolerant of competing ideas. (He was, after all, often proven right.) In his book, *Lawrence and Oppenheimer,* Nuell Pharr Davis described him thus:

> Alvarez most nearly carried the spirit of prewar Berkeley [where he had joined the faculty in 1938]. Like Lawrence, Alvarez characteristically wanted to hurry and do big things, and he too tended to treat other scientists as servants. But he showed much more positive readiness to cause pain. "I watched him identify negative protons several years prematurely with a cloud chamber," says a frequent visitor to Berkeley. "Then, years later, after the war I had lunch with him at the Faculty Club while he was building a multimillion-dollar bubble chamber. The project manager came by and Alvarez started chewing him out in front of the whole table. I left the table and came back and Alvarez was still at it. Seemed needless to me. As well as I could tell, not the poor guy's fault at all."
>
> Brilliant and capable, Alvarez divided his talent between experiments and public relations. "Frustrated," says another acquaintance. "He was the only one of the sunbathers who never got a Nobel prize [later rectified], and they kept him reminded. Fantastic ego. Loved to tell how amazed the groups he lectured to were at how young he was. He would brag so openly you would chuckle."

In one of the more controversial sequelae of the Manhattan Project, Alvarez became one of only five scientists to testify against their colleague and leader, Robert Oppenheimer, director of the project and the generally acknowledged father of the atomic bomb. Deeply concerned about the destructiveness of these weapons, Oppenheimer was reluctant to see the United States proceed with the far more destructive hydrogen bomb. Along with the four others, Alvarez testified that Oppenheimer was a security risk. All of Oppenheimer's other Los Alamos colleagues testified in his behalf. Nonetheless, the Commission found against him, pronouncing him unfit to serve his country any further. Most people then and now consider the decision a case of the end (getting rid of Oppenheimer's obstructionism) justifying the means (defaming his character).

How Alvarez came to his own conclusions in this case can at least be guessed at from another account in Davis's book:

> One of the leaders in the atomic establishment says that he
> was appalled by an intimation he caught in 1954 of the
> way anger and frustration had affected Alvarez' mind: "I
> remember a shocking conversation I had with Alvarez. It was
> before the Hearings. I want to make it clear that I am not
> giving his words but trying to reconstruct his reasoning.
> What he seemed to be saying was Oppenheimer and I often
> have the same facts on a question and come to opposing
> decisions—he to one, I to another. Oppenheimer has high
> intelligence. He can't be analyzing and interpreting the facts
> wrong. I have high intelligence. I can't be wrong. So with
> Oppenheimer it must be insincerity, bad faith—perhaps
> treason?"

Now that is an astounding piece of reasoning, leaving no room for disagreement, no room for alternative explanations of the same set of facts: only one road to truth. One could have predicted that this kind of mind would have some trouble getting along with geologists and paleontologists—low as they are in the scientific intellectual hierarchy, and so given to probing among a lot of unknowns in the quest for a consensus.

The political effects of this controversy did not stop at public obloquy among the principals, but trickled down (as it were) into the less public arenas of academic life. Younger scientists who saw a bandwagon in the impact hypothesis wanted to get on board. For any young scientist out of graduate school who is seeking an academic career, the plum job is a tenure-track position at a university. These are "hard money" positions, advertised in appropriate scientific journals, with a guarantee or promised guarantee of continued employment. In contrast, for "soft money" positions, there is no such guarantee but only an employment agreement on a year-to-year basis as funding permits.

In the past, tenure-track positions in Earth science departments, as elsewhere, became available through departmental expansion and normal turnover. But in the past decade or so, as a result of various cutbacks in academic funding in general and fewer opportunities for geology graduates in the commercial job market, such openings have become scarcer while the pool of potential applicants has remained large. It has been a buyer's market, so to speak, as well as a reverse bandwagon. Departmental review committees need to find reasons to reject applicants, and they have tended to avoid controversy and controversial people. If a job applicant has been involved in the K-T debate, his or her file is apt to be passed over. It is simpler that way.

One of the few safe refuges for younger members of the impact fraternity has been the Lunar and Planetary Institute in Houston, Texas. In recent years many such institutes have come into vogue: They are typically operated by a consortium of universities and funded by the federal government to carry on some specific kind of research that is of interest to Congress (or some other sponsoring agency) and to some of the faculty within the university consortium.

In this kind of institute, tenured faculty researchers make contributions to the institute's research program, but the actual employees of the institute have "soft money" positions. They are totally dependent on the waxing and waning of interest in their particular research activity, as are the overall fortunes of the institute itself. The sponsoring agency gains from such an arrangement by having no direct administrative or overhead responsibility for the institute. The universities gain advantage by having their faculty members expand their scientific efforts. It is the institute employees who face potential economic and career disadvantages; some institutes have been able to change their research directions as trends have dictated, while others have not been as fortunate.

The Lunar and Planetary Institute (LPI) was founded principally as custodian of the moon rock samples brought back by Apollo and preceding programs in the 1950s and 1960s, with a responsibility for disbursing the samples to other research installations. Beginning in 1970 and continuing to this day, the LPI also ran annual symposia on planetary matters. It has been funded chiefly by NASA, funding that has tended to go up and down over the years. No new moon rocks have been collected since the 1960s, and analysis of the existing samples was largely completed some time ago. So the LPI's continuing existence has occasionally come into question. Each time this has happened, however, those members of the academic community interested in planetary geology have helped keep it alive, in spite of the fact that new data from such planets as Venus and Mars has come in small amounts and at great expense.

But in 1980 along came the Alvarez hypothesis with its extraterrestrial perpetrator (and soon enough, perpetrators), along with the implication that it might happen again, this time to us. Almost immediately, the LPI intensified its studies of the effects of impacts on Earth. Another boon, in this sense, has been the recent cometary impacts on Jupiter. In any event, research scientists at the LPI and other institutes with similar funding that have hitched their wagons to the hypothesized K-T impact have in many ways the most at stake in this continuing debate: their jobs.

One way of keeping the debate going is to hold conferences at which the new hypothesis can be discussed and espoused. These conferences typically should be national or international and have the sponsorship of one or another august scientific body, preferably the National Academy of Sciences. It is important that contrary views be represented, but not too heavily. The first major impact hypothesis conference was held in 1981 in a resort complex at Snowbird, Utah. It was arranged principally by Kevin Burke, director of the LPI, and Eugene Shoemaker, head of the astrogeology branch of the U.S. Geological Survey (and the man who, with his colleagues, would later announce the impending arrival of the comet that struck Jupiter). A part of both men's jobs is to promote awareness of their fields of interest and to promote funding for them; both have been exemplary in such efforts, as well as making many scientific contributions.

Sponsorship of the National Academy of Sciences was secured for the Snowbird conference, but via what may seem an odd route. Among the Academy's various boards and committees, several are engaged in Earth sciences, but they were not involved in this conference. Instead, Academy support was gained through its Naval Studies Board, whose executive secretary, Lee Hunt, was an impact enthusiast. He and the Naval Studies Board staff were instrumental in raising funds for the Snowbird conference and making arrangements for the meeting itself—though one may legitimately wonder about the connection between dinosaurs and submarines.

In 1988 the same groups sponsored and held a similar meeting, Snowbird II, in the same Utah resort, but by the time planning began for a third meeting, called Snowbird III though held in Houston, the National Academy had decided not to sponsor it. Burke, Shoemaker, and others arranged that it be held under the auspices of the Lunar and Planetary Institute.

These several Snowbird conferences resembled other scientific conferences in many respects. They consisted of presentations, lasting from twenty to thirty minutes, by individual scientists on their research. Some of the speakers were invited; others had submitted abstracts that were then either accepted or rejected for presentation. The Snowbird conferences did not resemble symposia held by such scientific societies as the Geological Society of America, the American Association of Petroleum Geologists, the Paleontological Society, and the American Geophysical Union in that the emphasis at Snowbird was on impacts and their possible effects rather than on a broader and more evenly balanced presentation of opposing views.

Early in the impact controversy, a related matter caught the eye of America and the world. In 1983 readers of *Parade* magazine, that nearly ubiquitous Sunday newspaper supplement, learned from one of the magazine's regular contributors, astronomer Carl Sagan, that if nuclear war occured, it would visit upon the planet yet a further catastrophe beyond the destruction of targeted cities and the release of radioactive materials into nature. Sagan's theory, which was riveted into the public consciousness with the rubric "nuclear winter," maintained the aftereffects of such an outbreak would be just as lethal for people, plants, and animals in countries remote from the war zone as for the populations and ecosystems of belligerent regions.

The idea for the nuclear winter came directly from the meteorite impact theory and was published by Sagan and several colleagues in *Science* in 1983. On this paper Sagan based his startling piece in *Parade*. An all-out nuclear exchange, it announced, would produce an enormous quantity of smoke and chemical pollution from the burning cities, burning fuel depositories, even burning tires, and planetwide forest fires that would almost immediately ensue. In all, a hundred million tons of smoke would rise into the atmosphere and blanket the Earth, cutting off the Sun's radiation and producing freezing temperatures for at least several months, even in summer. Virtually no life could survive the freeze.

Nuclear winter immediately created widespread moral outrage internationally, strengthening the already great opposition to nuclear warfare at a time when it was still a dreaded option for the superpowers, as it had been since the late 1940s (and will continue to be in one form or another as long as nuclear weapons exist). As a political and moral idea, nuclear winter was a potent force for good, and many scientists were eager and happy to jump on the bandwagon in its support—even the many who had deep misgivings about the idea's scientific plausibility. It put many scientists, as physicist Freeman Dyson would later write in *Infinite in All Directions,* "in an awkward position." For it is the professional duty of a scientist to try and prove any hypothesis offered in his field wrong.

Dyson's own "take" on nuclear winter was that its existence would depend in great part on the nature of the soot that would be thrown up in the atmosphere. If the soot that enveloped the world were dry, Sagan's outcome might be right, since dry soot's optical properties would block heat from the Sun but allow heat from the

Earth to escape—both leading to a freeze. On the other hand, if the soot were wet, it would block heat in both directions, leading to quite a different outcome—essentially, no temperature change. In any case, Dyson himself chose what he called an "unheroic compromise." He kept his scientific doubts to himself until the facts became clear. It was, he said, "good to be honest but it is often better to remain silent."

Nevertheless, despite the nearly unfathomable complexity of the Earth's climate and the difficulty of making predictions about it, argument did finally surface, leading to an emotional and highly charged 1985 meeting of the American Association for the Advancement of Science. It was argued, for instance, that a nuclear exchange would send vast quantities of water vapor into the atmosphere, leading to a greenhouse (warming) effect that would override any cooling from the soot. Other scientists questioned the likelihood that global forest fires would break out. It would depend, they said, on whether the large fires caused by nuclear explosions would turn into firestorms, where a huge fire-caused updraft sucks in ground-level air at hurricane-wind velocity, feeding the fire and fanning its flames into what is almost a self-sustaining phenomenon. The atomic bomb at Nagasaki caused no firestorm; the one at Hiroshima did, burning out the core of the city. Just when a fire will become a firestorm remains unknown.

More detailed climate models suggested that the nuclear winter effects would be fewer overall, or less long-lasting, or less global than Sagan predicted. Its proponents were praised for bringing the entire matter up and for adding another arrow to the quiver of those who believed nuclear weaponry was unusable; but they were also called "totally irresponsible" for stressing the dire effects without also stressing the uncertainties inherent in such modeling, thus running the risk of discrediting the entire field of climatology. In any event, over time, most students of this phenomenon have concluded that the most likely outcome would be not a nuclear winter but a far less devastating nuclear autumn—by no means diminishing the obvious need to avoid nuclear war.

What is of interest here is the fact that a wholly new hypothesis—about nuclear winter—was so quickly inspired by another hypothesis (meteoritic impact at the K-T boundary) before the latter had been run through the usual scientific gauntlet of *its* peers. Of course, the nuclear winter hypothesis could be judged on its own scientific merits, regardless of any meteorite theory. But the publicly unquestioned (and, as far as many scientists in other fields knew, unchallenged) assertion that a meteorite had once caused a similar catastrophe lent early credence to nuclear winter. Even more strangely, perhaps, people in the Alvarez camp took to suggesting that early opponents of their hypothesis were willy-nilly undermining the nuclear winter hypothesis and thus lending comfort to warmongers! In one remarkable epistolary squabble, Alvarez accused Robert Jastrow of just this, pointing out darkly that Jastrow was then also a supporter of Star Wars research: QED. Thus did a man who had accompanied the first atomic weapon ever detonated in anger to its target in Japan accuse another of being a militarist.

Such was one impact of the impact theory. In another realm—an esoteric biological debate on the exact nature of the evolutionary process—it played a supportive role for one side, even though it was, itself unproven in any normal scientific sense. This debate, which still continues, involves the standard Darwinian model of evolution. This widely accepted model speaks of a generally steady and gradual process by which the force of natural selection constantly fine-tunes populations of creatures to fit into changing environments, thus over time giving rise to new species that can adapt to new circumstances.

Those who oppose the Darwinian model propose instead a theory of "punctuated equilibrium," wherein species remain relatively constant over long periods of time, then suddenly undergo periods of rapid biological creativity, leading to new arrays of species in short periods of time. It is not at all surprising that such theorists looked upon the meteorite impact theory with favor. Again, the notion of punctuated equilibrium can and will stand or fall on its own merits, regardless of the fate of the impact hypothesis. But proof that the impact had the effects it is alleged to have had would make punctuated equilibrium all the more plausible, supplying a mechanism for bringing about creative crises in evolutionary history.

Other remarkable hypotheses would soon arise, directly tied to the impact hypothesis, and these too would have widespread impact themselves—if briefly. The predicted "silly season" arrived, as well as numerous matters of rigorous and not-so-rigorous science. We shall return to this aspect of the debate in subsequent chapters, but first we must look more thoroughly into the activities of another major player in the controversy: the media.

SEVEN
Media Science

SCIENTISTS FOLLOW A RELATIVELY STANDARD
procedure to achieve *scientific publication,* which means, basically, a
formal announcement of research findings. This standard procedure is one that at times seems, even to scientists, a bit tedious.
Nevertheless, for the most part, it has worked over time to everyone's advantage. Generally the procedure is as follows:

An investigator writes up his or her findings and submits the
manuscript to a scientific journal for publication. Usually the journal
is associated with a relevant scientific society, and the editor is a
respected figure in the field who has taken time off from his own
research to do the editorial work.

Typically, the editor then sends the manuscript to a group of
the author's peers for review. Particularly on controversial or inno-
vative subjects, which are often the most difficult to judge, the editor
seeks reviews by individuals with viewpoints on various sides of the
topic. On the basis largely of these reviews, the editor will accept the
manuscript as is, ask for revisions, or reject it altogether. A rejection

is usually based on incompleteness of the investigation, faults in the manuscript's presentation, or lack of consideration of studies that reach contrary conclusions.

The system works pretty well, invoking peers to act not only as judges of one's work but, in a sense, as stewards of science and scientific method in general. Even so, the system has flaws, as any human system does. But one flaw—especially today, when the amount of scientific investigation in a variety of fields is growing rapidly, almost exponentially—has to do with time. Most scientific (that is to say, technical) journals appear monthly or quarterly. Given the time needed to review an article and make revisions, as well as the large backlog of manuscripts awaiting publication, six months or even several years may elapse between a manuscript's submission and publication. By that time, it may be old hat.

Investigators often alleviate the time problem in making their work public by making formal presentations and holding informal discussions at scientific meetings and conventions, and by using internal networks of communication among scientists in a particular discipline. Such networks are now served, even more rapidly, by electronic means. In all of these modes of communication, an investigator's peers can exercise one or another degree of internal expert judgment, even though the presentation of data and conclusions is more informal and more rapid than in the traditional publication mode.

Investigators also have at least two other ways of presenting their findings. One is to go directly to the public via the popular press, bypassing conventional technical journals, one's peers, and all the other procedures that have arisen to make science self-corrective and responsible. Usually findings presented this way have to do with a dire threat to mankind or to the environment—such as an imminent earthquake—or a remarkable cure that has so far utterly eluded medical researchers. Few if any legitimate scientists ever take this route, and no important or lasting contribution to science has ever been reported to the world by this foreshortened means. It can properly be called tabloid science.

On the other hand, some routes to scientific publication that sidestep the problematical matter of time to some degree are nonetheless accepted. Publication can be accomplished in one of the more general and widely disseminated (though still technical) journals such as Great Britain's *Nature,* or *Science,* the journal of the American Association for the Advancement of Science, a Washington, D.C.–based organization of long standing that, as its name suggests, fosters the progress of science in its many disciplines.

Both *Nature* and *Science* appear weekly, and insiders sometimes refer to publication in them as a "quickie." Of practical necessity, the review procedures are more cursory, but the expectation is that a brief report in their pages will be followed in due course by a more lengthy and complete account in a more conventional scientific journal devoted entirely to the particular discipline. But sometimes this does not occur: the "quickie" publication becomes an end in itself. For example, Carl Sagan and some colleagues published their concept of nuclear winter in *Science.* Then, before the matter was taken up in the technical journals of the various scientific disciplines involved, it appeared in *Parade.* As we have seen, when it eventually came under the formal scrutiny of other scientists, it was found to be, at best, overstated as a scientific finding (regardless of its political and moral value). But by then it was too late to correct whatever impression had formed in the public mind about its actual scientific validity. This, as we will see, was largely the case with the Alvarez hypothesis and the ensuing debate about the mass extinction at the K-T boundary.

The two weekly journals also act as sources of communication between the various scientific disciplines—for example, allowing astronomers to keep up with developments in molecular genetics, if they wish. They also act as intermediaries between the scientific community and the public. Not many members of the public do, or even can, read *Science,* however. Even its reports on scientific developments, let alone its technical presentations by scientists, are over the heads of most citizens. Its presentations are disseminated to the public largely via science journalists—newspaper and television reporters assigned, often temporarily, to the science beat, and science editors on newspapers, television, and in the magazine industry—who use *Science* as a source.

To aid these journalists, *Science* and its British counterpart *Nature* often send out press releases about the articles they consider of potential interest to the public. It is always a matter of judgment to decide what the public reasonably ought to know and what those other intermediaries—popular editors and journalists—think the public will find palatable. Of course, public attention tends to light upon what is relatively quick, simple, and sensational, rather than on what seems overly long, complex, and tedious. All editors and reporters know that and have to live with it. So to be an intermediary anywhere in the region between scientific information and the popular mind is to walk a tightrope between the often opposing needs of education and entertainment. And therein lies a great deal of room for irresponsibility. While scientists and others properly decry the popular press's frequent lapses into irresponsibility, such lapses *within* the scientific community are all the more deplorable.

In the late 1960s and early 1970s, the loud controversy over polywater was just such an embarrassing lapse, and since it has numerous parallels with the controversy surrounding the Alvarez hypothesis, it is worth recalling in some detail.

The polywater dispute was a scientific cause célèbre. Polywater was supposed to be a new form of water with wonderful properties of potentially great benefit to mankind. It was later found that the whole thing amounted simply to contamination in the laboratory test apparatuses.

The dispute has been well chronicled by Felix Franks, a senior research fellow at Cambridge University, in his book appropriately entitled *Polywater*. It was played out largely on the pages of *Science* and *Nature*. Most of the overabundant articles in these two journals presented information on the existence and properties of polywater, extolling its virtues, with only a few challenging or disclaiming the whole thing. These "quickie" publications were in some combination incomplete, incorrect, inaccurate, overinterpreted, and hastily written. The science news reporting community, by and large, took the articles and associated comments and news releases as fact and broadcast the story widely to the general public.

It had all started in the 1960s, when studies by Soviet investigators indicated that they had found a new and anomalous form of water. The fluid was produced from ordinary water, but it did not freeze or boil like water, and it hardly resembled water in any of its properties. The investigators presented their results at scientific meetings in the West, at the time when Sputnik had raised concern that Western science was falling behind Soviet science. Here was, perhaps, yet one more example.

The next step up, and a giant one it was, was published in *Science* in 1969 by a respected American spectroscopist. His results, he stated, demonstrated that an anomalous form of water did indeed exist. He coined the term *polywater* for it. That was news. (A few years later that same individual recanted on his earlier pronouncements. That was not news.)

The final step came in the pages of *Nature,* also in 1969, with the publication of the following concluding remarks from an American investigator:

I need not spell out in detail the consequences if the polymer phase can grow at the expense of normal water under any conditions found in the environment. Polywater may or may not be the secret of Venus's missing water. The polymerization of Earth's water would turn her into a reasonable facsimile of Venus.

After being convinced of the existence of polywater, I am not easily persuaded that it is not dangerous. The consequences of being wrong about this matter are so serious that only possible evidence that there is no danger would be acceptable. Only the existence of natural (ambient) mechanisms which depolymerize the material would prove its safety. Until such mechanisms are known to exist, I regard this polymer as the *most dangerous material on earth.*

Every effort must be made to establish the absolute safety of the material before it is commercially produced. Once polymer nuclei become dispersed in the soil it will be too late to do anything. Even as I write there are undoubtedly scores of groups preparing polywater.

Scientists everywhere must be alerted to the need for extreme caution in the disposal of polywater. Treat it is as the most deadly virus until its safety is established.

Shades of *The Andromeda Strain*!

There you are: a new and potentially very dangerous substance had been taken as bona fide because it was reported in *Science* and *Nature*. Throughout the whole dispute the exact physical properties—both good and bad—of this presumably new and anomalous form of water always remained somewhat of a mystery. The reader may well wonder how the dispute ever reached the fever pitch that it briefly did; after all, water is water, and that's that.

But when the bandwagon came along, everyone wanted to get into the act. Many scientists were taken in by the publicity campaign and the news articles. Polywater was where the action was. A surprisingly large number of scientists dropped whatever research they were doing to study this presumably more important substance. The number of articles in the United States scientific press increased from two in 1966 to fifty in 1970, before falling back to two in 1974, when it became clear that polywater simply wasn't. This whole incident bordered on a science gone amok— a *pathological science,* as Franks refers to it.

The whole thing fed on itself. Items such as the following appeared in newspapers and newsmagazines: "Polywater might be used in the development of a steam automobile, as a moderating fluid in nuclear reactors, as a superlubricant, and as a corrosion inhibitor." One national newsmagazine wrote that by now only "a few skeptics believe polywater is actually a mixture of real water and the material of the capillary tube, but many scientists are prepared to accept it as a truly new form of water." A number of the more extroverted scientists gave unsolicited as well as solicited interviews to the media—for their own self-adulation or for whatever other reasons.

In the science community the general atmosphere became polarized. You were either a believer in polywater or a rejectionist— and no quarter was given. Some impugned the professionalism and motives of others who had a contrary view. The debate was carried in the public press, with inappropriate and regrettable remarks by some scientists. As doubt began to creep in as to whether polywater really existed, those manning the polywater battlements stood firm. In *The New York Times* appeared a statement that "despite the new findings, a check of eminent scientists who believe in the existence of polywater showed no weakening of their convictions."

It all ended rather quickly, and we conclude with the following three summary statements by Felix Franks on various aspects of the polywater debacle.

> The history of polywater provides clear evidence of questionable publication practices. One of its distinguishing features is the proliferation of "quickie" publications, for instance in *Nature* and *Science.* The majority of those who got involved in the affair published no more than one such letter in one of these journals. The pressure was great—in the general rush for priority there was not enough time to perform enough experiments and think about the results before committing them to paper. . . . One can only speculate either that the refereeing was superficial or that on occasions the editors overrode the recommendations of referees in order to promote further discussion, or to provide a forum for unorthodox views.

> That self-deception was one of the ingredients of the polywater affair, especially after the bandwagon had begun to roll, cannot be denied. Therefore, on this count, the charge of pathological science must be taken seriously.

> Without doubt, the most powerful single element responsible for turning polywater research into pathological science was the involvement of the mass media, which thrive on dispute and confrontation.

To return to the meteorite impact story: *Science* magazine became the publication medium of choice for those whose work supported the Alvarez hypothesis. From 1980 until the mid-1990s, *Science* appears to have had a considerable bias in favor of the impact hypothesis—although its editors insist otherwise. They did, to be sure, publish some papers in opposition to the hypothesis, and some of these have been important. But the facts, quantitatively, speak for themselves. Between 1991 and 1993, for example, *Science* published eleven articles favorable to the hypothesis and two unfavorable. It could be said that this observation is merely sour grapes on the part of the hypothesis's opponents, but before long the bias was so evident to members of the Earth science community that few even bothered to submit to *Science* a manuscript that espoused a terrestrial cause for the K-T extinction events.

The bias became a matter of public discussion as early as 1985. In a report about the annual meeting of the Society of Vertebrate Paleontologists (the dinosaur experts) on October 29, 1985, reporter Malcolm Browne of *The New York Times* noted a general agreement among the participants that the impact theory had benefited paleontology by pushing scientists to test the hypothesis and pursue new lines of research. But he also quoted one participant to the effect that it had also polarized the scientific community so severely that scientific careers were at stake. This statement, Browne said, was supported by many other paleontologists, who also asserted that some scientific journals, notably among them *Science,* favored the impact theory in their acceptance or rejection of papers for publication.

Browne went on to quote *Science*'s editor, Donald E. Koshland, Jr., as saying: "We bend over backward to be fair in our selection of the reviewers who decide whether a paper will be accepted or not. Contributors should realize that because we receive so many fine papers we're obliged to reject four out of five submissions, and these include a lot of very good papers that are published elsewhere. We only have room for what we deem to be the best."

Seven years later, in 1992, Koshland was quoted in *The Chronicle of Higher Education* as saying: "We've been very careful. We've published arguments from both sides, and we've published criticism from both sides. There's no bias." For this article, reporter Kim McDonald talked also with R. Brooks Hanson, a senior editor at *Science* responsible for the publication of papers in geology. Hanson said the magazine has actually rejected "far more papers that support the impact theory than those refuting it, because more of the former were submitted." McDonald's article went on as follows:

> Gerta Keller, a professor of geology at Princeton Uniersity, says she is one of those who have been rebuffed. She says she submitted a paper two years ago to *Science* that showed that a dust cloud from an asteroid impact could not have existed, because many marine plants that require uninterrupted sunlight were unaffected. Instead of sending her paper to outside reviewers, she says, the journal's editors returned it with a note that stated, "This is not of interest to the public." Neither Mr. Koshland nor Mr. Hanson recalls the paper.
>
> Ms. Keller, who says her colleagues have had similar experiences with *Science*—and, to a lesser extent, *Nature*— notes that the influence of these journals has made it increasingly difficult to oppose the asteroid hypothesis, get grants to finance work that might contradict the idea, and persuade young scientists to work on opposing ideas.
>
> "The perception is that whatever gets into these popular magazines is the truth," she says.
>
> "I'm trying to get people to look at these things, and it's virtually impossible," she adds. "If they are on the chalk mark for tenure, they would not have the ego to take the enormously berating attacks against them."
>
> "I don't think it is an organized conspiracy," Ms. Keller says. "But so many people have jumped on the bandwagon and so many reputations are at stake, they have lost the ability to be objective."

Others, less sanguine, have pointed to a rather strong connection at *Science.* Koshland was from Berkeley, as was Alvarez. Koshland's predecessor, the distinguished Philip Abelson, was also from Berkeley. For a time in the late 1980s, another Berkeley product, Alvin Trivelpiece, served as publisher of *Science.* One can make of this what one wishes, but it is important to point out that still another Berkeley product, William Clemens, called the Alvarez hypothesis "codswallop" and in turn got his head handed to him by Alvarez.

Arguments will no doubt continue (and, no doubt, will not reach any firm conclusion) about what if any bias operated in *Science*'s publication of technical reports and scientific papers on the impact hypothesis. But a clear tendency, if not an outright bias, is unmistakable in another part of the magazine—its news reports. Here staff reporters provide weekly accounts of various developments in many fields of science. For most of its lifetime, the impact controversy has been reported there by Richard Kerr.

What follows are some of the headlines and introductory material from the magazine's reports from 1980 to 1994 on this controversy.

OCTOBER 31, 1980:

Asteroid Theory of Extinctions Strengthened

An asteroid may have hit Earth at the close of the dinosaur age, but how that impact may have affected life is still obscure.

NOVEMBER 20, 1981:

Impact Looks Real, The Catastrophe Smaller

Diverse specialists now agree that the evidence for a huge asteroid (or comet) impact is impressive, but they have scaled down its effects.

SEPTEMBER 2, 1983:

Extinctions and the History of Life

Now that, for many at least, asteroid impact has been accepted as a causative agent in mass extinction, attention turns to the wider view.

NOVEMBER 11, 1983:

Isotopes Add Support for Asteroid Impact

Osmium isotope analysis supports an asteroid impact 65 million years ago but cannot exclude a huge volcanic eruption.

MAY 8, 1987:

Asteroid Impact Gets More Support

The global distribution of shocked quartz at Cretaceous-Tertiary boundary argues for an asteroid or comet impact and against a volcano as a cause of the mass extinction.

AUGUST 21, 1987:

Searching Land and Sea for the Dinosaur Killer

The impact that triggered a mass extinction and possibly the death of the dinosaurs left clues to its location.

FEBRUARY 12, 1988:

Was There a Prelude to the Dinosaurs' Demise?

Something seems to have been going on just before the geologic instant of the Cretaceous-Tertiary boundary and its large impact.

NOVEMBER 11, 1988:

Huge Impact is Favored K-T Boundary Killer

A large impact rather than a volcano is widely taken to be the primary agent of destruction at the end of the dinosaur age.

DECEMBER 9, 1988:

Snowbird II: Clues to Earth's Impact History

At the 1981 conference at Snowbird, Utah, on the effects of large impacts, it became clear that an asteroid impact was not as outrageous an explanation for the mass extinction at the end of the age of dinosaurs as it had seemed. At Snowbird II this October, the large impact hypothesis prevailed over cataclysmic volcanic eruptions as the cause of mass extinctions 66 million years ago. . . . But even accepting that conclusion, which everyone does not, a wealth of questions remain [sic]. To answer some of them, researchers of all sorts are having to look in unprecedented detail at the geologic record. From the new detail the answers are beginning to come.

JANUARY 10, 1992:

Extinction by a One-Two Comet Punch

As if one huge impact 65 million years ago weren't bad enough, two or more blows may have ravaged life.

AUGUST 14, 1992:

Huge Impact Tied to Mass Extinction

Radioisotopic dating now has forged the final link between the immense crater in Yucatan and extinctions 65 million years ago, when dinosaurs disappeared and our ancestors began to flourish.

JUNE 5, 1992:

Did an Asteroid Leave Its Mark in Montana Bones?

MARCH 12, 1993:

Second Crater Points to Killer Comets

SEPTEMBER 17, 1993:

How Lethal Was the K-T Impact?

The asteroid that hit Earth 65 million years ago appears bigger than previously thought, but scientists have new doubts about its ability to kill the dinosaurs.

MARCH 11, 1994:

Testing an Ancient Impact's Punch

Did the impact at the end of the dinosaur age deliver a haymaker to life on Earth? Results newly reported from a "blind test" of the marine fossil record suggest it did.

No one reading these successive headlines and teasers could come to any other conclusion but that an impact of some sort definitely occurred at the K-T boundary and wiped out the dinosaurs and other forms of life. An attentive reader might get the sense that there was *some* disagreement, but only on the part of a very small minority of scientists. (A careful reader of just these headlines might also find a certain amount of internal confusion in them. Was it one asteroid or more? Was it a comet? Or several? If it was a comet, then why was it later referred to as an asteroid? Did it [or they] strike 65 or 66 million years ago?)

Such objections could be called quibbling, but a certain inconsistency in regard to facts was evidently arising. In any event, there can be no doubt about where the *reporters* for *Science* stood. The overwhelming preponderance of headlines alone implied or said outright that the impact extinction theory was a done deal.

Even so, Richard Kerr did not feel that the message had gotten across. In a 1989 article in *The Washington Post* he wrote: "The public . . . has not gotten the message yet. Leaning over backwards to appear even-handed, the news media portray opposing camps of equally credible scientists slugging it out over whether the catastrophe was the aftermath of a large impact or solely a huge volcanic eruption. In reality, volcano advocates remain a tiny minority and the evidence is solidly on the side of an impact."

It is difficult to imagine what Kerr was thinking of when he bemoaned public perception of the impact hypothesis. To this day, most popular science magazines and most newspapers routinely refer to the asteroid impact and its resulting extinctions as facts. But what of the scientific community?

Kerr's mention of a tiny minority of scientists opposing the hypothesis calls to mind a favorite source of information used by journalists: opinion polls. One may question the validity of these polls on many grounds, and surely they measure only some body of opinion at a particular time—as opposed to the actual truth of any matter. But two polls have been conducted among scientists on the impact hypothesis. As we mentioned, Malcolm Browne's informal poll among paleontologists in 1985 found that while many of them were willing to go along with the idea of an impact, only four percent thought that an impact could have had anything to do with the mass extinctions. So Richard Kerr's "tiny minority" would not seem to apply to paleontologists—at least not to those attending their 1985 meeting. What of other scientists in the controversy?

The other poll was conducted in 1984 and reported in the December 1985 issue of *Geology* by Antoni Hoffman of the Polish

Academy of Sciences and Matthew Nitecki of Chicago's Field Museum of Natural History. The figures for those accepting the Alvarez hypothesis that an impact caused the mass extinction are as follows: 9 percent of British paleontologists, 10 percent of Soviet geoscientists, 14 percent of German paleontologists, 16 percent of Polish geoscientists, and 31 percent of American geophysicists. Even among American geophysicists the figure is quite low—a minority in all cases—meaning that a *majority* (and a large one) of scientists opposed the Alvarez hypothesis.

Evidently, Richard Kerr was not aware of these surveys. In any event, they do represent a disturbing schism between the public's view of what was going on and the actual opinions within the Earth science community. That *Science* played a role in creating this schism is unarguable.

Not all press coverage of this controversy has been one-sided, however. In several instances, and most notably on the editorial page of *The New York Times,* there have been more even-handed statements. On November 2, 1985, for example, the *Times* editorialized thus:

Dead Dinosaurs, Live Politics

A strange seed has taken root among paleontologists, the scientists who study fossil life forms. Fear is a bizarre phenomenon to find in 20th-century American science, as well as a ridiculous way to resolve scientific controversies. Yet some paleontologists fear their careers may be impeded if they oppose the currently fashionable theory that a meteorite collided with Earth and caused the extinction of the dinosaurs some 65 million years ago.

The dispute has so divided the scientific community, Keith Rigby, a paleontologist at the University of Notre Dame, told Malcolm Browne of *The New York Times,* "that acceptance or rejection of grant proposals and papers may depend on the personal views of the reviewers assigned to pass on them. Scientific careers are at stake."

One reason for the unusual heat is that the extinction idea is more than a hypothesis about the past: it has a political present—the nuclear winter conjecture, which has become a major talking point of the nuclear freeze movement. The dinosaurs are supposed to have perished because the dust clouds kicked up by the putative meteorite blotted out the sun. That idea, put forward in 1980, led others to propose that the smoke from burning cities in a nuclear exchange would block the sun's rays, producing extreme cold and darkness for months and sending humans on the way of dinosaurs.

The validity of the two hypotheses should not be linked: they will stand or fall separately. Nevertheless, a few paleontologists who dispute the dinosaur extinction theory have found themselves being branded as militarists, on the grounds that their skepticism undermines the nuclear winter thesis.

Another cause of friction is the clash between academic disciplines. Many proponents of the meteorite-extinction thesis are physicists, boldly trespassing on the paleontologists' turf. Such trespassers in science often see jewels unnoticed by the resident experts. The intrusion aside, paleontologists feel that the physicists look down on them. Many also feel that the meteorite hypothesis is simply wrong. The extinction of the dinosaurs, they say, took place over millions of years and cannot be explained by a single catastrophe.

The meteorite-extinction and nuclear winter ideas have provoked much useful research, and both make vivid reading for the public, but neither yet enjoys a solidity matched by its publicity. A committee of the National Academy of Sciences recently concluded that nuclear winter is a clear possibility, but that no precise measures of its extent can yet be established. If the physicists and paleontologists can set aside their passions and personal stakes in rival theories, progress on both hypotheses will be quicker.

Passions have been slow to cool, but progress has been made nonetheless. But interestingly, in the entire course of this debate, no such reasoned call was ever forthcoming from the editors of *Science*.

As paleontologists and other Earth scientists working diligently in the field found one piece of evidence after another that the K-T extinctions were not an instantaneous event, as the impact hypothesis predicted, their frustration and anger were understandable. Their results never reached the public. Instead, there seemed to be a direct pipeline from the news reports in *Science* to science journalists to articles in newspapers and reports on television and radio: all to the effect that there had been an impact and that it did in the dinosaurs.

One of the few avenues of recourse open to these scientists was to write a corrective letter to *Science* and hope that *Science* would publish it. The report in the March 11, 1994, issue of *Science* on the Snowbird III meeting and the "blind test" on the marine fossil record in Tunisia that Gerta Keller had studied prompted such a reply. In a "blind test," samples from a given geologic section are supplied to several investigators for analysis without identification as to their stratigraphic level: it is a reasonable approach to verify scientific findings. Keller's corrective letter is as follows:

> In the recent issue of *Science* vol. 263, March 11, 1994, in "Research News" p. 1371–1372, Mr. Richard Kerr reported on a recent meeting on Impacts and Catastrophes in Houston. He specifically reported on a "blind test" of planktic foraminiferal extinctions at El Kef and separately on a field trip to the K/T boundary sections in northeastern Mexico. Both reports are fraught with numerous factual errors and misrepresentations as indicated below. I hope that the following letter correcting these factual errors will be published in a subsequent issue of *Science*.
>
> 1. Keller's initial El Kef report was published in 1988, not 1989.
>
> 2. Keller's 1988 paper reported the disappearance of 14 species or 26% of the species below the K/T boundary and not 29%.
>
> 3. Smit never published any species census data on El Kef, thus he has produced no evidence either for or against species

extinctions prior to the K/T boundary, contrary to Kerr's implications.

4. Robert Ginsburg did not collect the new samples for the blind test. In fact, he has never been to El Kef.

5. Smit did not "minimize the influence of rare or misidentified species" by combining all four blind test results. In fact, he only extracted a total of seven out of sixty-two species that disagreed with my 1988 paper in support of his claim of no evidence for pre-impact extinctions.

6. Contrary to Kerr's report, the purpose of the "blind test" was to test Smit's (1982) extinction model of all but one species extinct vs. Keller's (1988) model with about 1/2 extinct at the K/T boundary, 1/3 species surviving, and the remainder extinct below the boundary.

7. Why did Kerr not report that all blind test investigators reported between 36% and 46% of Cretaceous taxa ranging into the Tertiary?

8. It was Keller (1993) who could not confirm Brian Huber's (1991) study rather than the reverse as implied by Kerr. Huber's comments are therefore hardly those of an objective critique.

9. Kerr is incorrect in stating that while paleontologists discussed the findings of the blind test, sedimentologists worried about deposition. The blind test was discussed only on the last day of the meeting and many more paleontologists than sedimentologists participated in the field trip and the discussion on deposition.

10. The pre-conference field trip was not organized by Robert Ginsburg, but by Wolfgang Stinnesbeck from the Universidad Autonoma de Nuevo Leon, and Gerta Keller from Princeton University.

11. The impact tsunami scenario did not win the day as Kerr states. Quite the contrary.

12. Contrary to Kerr's statement, sedimentologists on the field trip found no evidence for up-and-downhill currents.

13. Kerr completely misrepresents the sedimentologists by quoting Robert Dott as speaking for all by concluding that it

was an impact-induced tsunami deposit. In fact, this statement was made at the Houston meeting with Kerr present and was immediately countered by sedimentologist Don Lowe from Stanford University who, speaking for the majority of the field trip participants, concluded that the deposit was complex and represented multi-event deposition whose origin could not be determined without further field studies. Why did Kerr ignore this majority view in his report?

These many factual errors do not inspire confidence in Richard Kerr's understanding of the controversy.

Science did publish Keller's letter in an edited form.

In another letter, William Zinsmeister commented in a more general fashion on the same meeting.

For years, the Research News articles have been one of my favorite sections in *Science.* In the past, the Research News section has presented unbiased summaries of new ideas and recent research discoveries. Unfortunately, these high standards that *Science* has maintained over the years have declined in the reporting of the K/T extinction events. This is a clear case of the loss of scientific objectivity by the reporter, in this case, Mr. Richard Kerr. This loss of objectivity was clearly displayed in his most recent Research News article, "Testing an Ancient Impact's Punch," summarizing the Houston Impact and Catastrophe meeting. To read Mr. Kerr's article, one would believe that the "blind test" and "tsunami deposits" were the only topics covered at Snowbird III. Unfortunately for the scientific community, Mr. Kerr failed to note that a number of papers were presented addressing important questions, such as—1. Was the decline in biodiversity during the last 10 millions [of years] of the Cretaceous related to, or separate from the K/T extinction? 2. Why did some important groups of marine invertebrates disappear in the high latitudes prior to their disappearance in the low latitudes during the Cretaceous? 3. If the impact event was as catastrophic as portrayed, why is there no geomorphic signature in the rock record or layer of burnt and twisted dinosaur bones? These are fundamental questions that have to be addressed before we will have a true understanding of events at the end of the Cretaceous.

A number of these questions were discussed at a K/T symposium several months ago during the annual Geological Society of America meeting in Boston which Mr. Kerr attended, but did not report. Why were these important papers never mentioned by Mr. Kerr? Could it be because Mr. Kerr has lost his scientific objectivity?

Mr. Kerr's lack of objectivity is clearly reflected in his habit of interviewing only those who support the impact hypothesis. In the few cases where he has interviewed scientists who have questioned some aspects of the K/T extinction, their comments for some reason are never reported. The only exception is Dr. Gerta Keller where Mr. Kerr emphasizes only the negative aspects of her ideas and cites only those scientists that disagree with her. A science reporter has an obligation to interview all the participants in a scientific controversy and to report their comments in a fair and unbiased manner.

Snowbird III was the first impact meeting that I have had the opportunity to attend. The dynamic and enthusiastic discussions during the meeting were stimulating and, at times, entertaining. It is unfortunate that Mr. Kerr did not see fit to share the excitement and many new ideas and data with the scientific community.

Science has an obligation to the scientific community to report all aspects of any controversy in a fair and unbiased manner. In the future, *Science* should assign a reporter to K/T extinction meetings who has the ability to maintain objectivity concerning this important scientific issue. For *Science* to allow Mr. Kerr to continue his biased reporting of the K/T extinction controversy would be a disservice to the tradition and high standards that *Science* has maintained over the years.

His letter was not published by *Science.*

In still another letter to *Science,* this time on the reporting of the Snowbird II meeting, Alan Rice of Rhodes University in South Africa commented as follows: "Kerr confided to me at the Geological Society of America meeting in Phoenix last year that dissenting views would not find airing in *Science.* Such policy is certainly in place in his review of Snowbird II." Rice's letter was published in *Science* in an edited form and without this comment.

EIGHT
Iridium and Shocked Quartz

THE MOST IMPORTANT PIECE OF GEOLOGIC evidence to which the Alvarez camp could point in support of the meteorite impact was the presence of the high-iridium layer in the claybeds of Gubbio and several other places around the globe, and the presence here and there of minerals—notably quartz—that were shocked into unusual forms, presumably by the impact.

Both of these phenomena in fact amounted to, circumstantial evidence at best, but in the early 1980s it was not unreasonable to find them compelling evidence pointing to a meteorite. Since that time, however, the geologic research that has taken place on both phenomena—some of it impelled by the Alvarez hypothesis—has not been at all kind to the impact theory.

As noted in Chapter 1, iridium does not occur in great concentrations in the Earth's crust, but there are a few places where it does occur in mineable quantities. In Siberia it comes from the Noril'sk deposits, part of which is called the Siberian Traps—an area that, like the Deccan Traps, is the result of intense volcanism long ago. In South Africa iridium has been mined from the Bushveld Complex, a sequence of intrusive rocks that also originated from deep within the Earth's interior. Elsewhere, relatively high concentrations of iridium are found in certain mineral deposits—South Africa's Bon Accord nickel-iron deposit, for one—and under certain conditions iridium with values of 2 to 20 parts per billion (ppb) has been found in association with the formation of bituminous coals. Nevertheless, Alvarez's suggestion that the K-T claybeds with 1 to 10 ppb of iridium were of meteoritic origin was a reasonable enough assumption in 1980. Only three years later, however, new findings might well have given the impactors pause.

These findings were from Hawaii and came about as a result of research carried out, for purposes unrelated to impacts and dinosaurs, by William Zoller, a chemist at the University of Washington, and colleagues in 1983. The volcano Kilauea is a *hot spot* volcano. Underneath the tectonic plate on which the Hawaiian Islands rest lies what might be thought of as a live plume of molten material rising from the Earth's core to the surface. The plume is stationary, and as the oceanic plate moves over it, it periodically punches up through, creating a volcanic island. The Hawaiian islands were all caused by the same plume: as their plate moved northeastward, they became quiescent and began to erode away. Kilauea, on the island of Hawaii, is simply the latest expression of the underlying plume. It erupted in January 1983.

On behalf of the National Oceanic and Atmospheric Administration's program of Global Monitoring and Climate Change, the Zoller team collected atmospheric particulate matter from the eruption on air filters. Much to their surprise, and that of just about everyone else, they found high levels of iridium in the particulates, along with other elements including antimony, arsenic, copper, zinc, gold, selenium, mercury, and cadmium. The iridium in the fine airborne particles constituted 630 ppb, which is a whopping *11,500 times* the concentration of iridium in Hawaii's basaltic rocks (which is 0.055 ppb).

This was one of the most important findings in the entire K-T debate. Meteorite impacts were thought to produce high iridium concentrations, but now it was clear that volcanoes were unquestionably responsible. The volcanic concentrations, however, were neither in the basaltic flows nor in the associated ashes. They were in the airborne particles that, at least in large eruptions, are carried up into the stratosphere and distributed globally. Interestingly, this was just what had been found in K-T sections where the iridium was associated with fine inorganic clay particles and *not* with the calcium carbonate plankton remains in the limestones. To its credit, *Science* promptly published Zoller's results.

The Kilauea eruption also produced a high fluorine concentration, and Zoller surmised that fluorine was partly responsible for the high concentration of iridium in the atmospheric particulates, especially since in the presence of fluorine, iridium occurs in the form of the highly volatile chemical compound iridium fluoride. The next logical question to be investigated was: Could iridium in this volatile form actually be deposited around the world and become part of the geologic record? The find in Hawaii suggested it was possible but didn't prove it had ever actually happened.

Affirmative answers soon came, however, from Kamchatka Peninsula, the northeasternmost part of Asia adjacent to the Aleutian Islands. There Russian scientists S. B. Felitsyn and P. A. Vaganov investigated the iridium concentrations in fine volcanic ash particles from five separate eruptions that had previously occurred there.

The volcanoes on Kamchatka are of the subduction type, meaning that they occur where one tectonic plate plunges under another. Such volcanoes produce magma that is acidic, as opposed to the more basic magmas produced by hot spot volcanoes and at midocean ridges, where plates are splitting apart. The eruptions studied on Kamchatka had occurred between 7,500 years ago and the present. For each one the iridium concentrations were found to be 1 to 4 ppb—comparable to the K-T sections.

An important finding here was that the smaller the particles—ranging from 1 millimeter to 0.1 millimeter—the greater the concentration of iridium. This suggested that in its volatile form, iridium has a "preference" for collecting on finer particles (which have a higher ratio of surface area to volume).

Now, in a collection of particles—both fine and coarse—that are being carried off in the atmosphere from a volcanic eruption, one would logically expect the coarser ones (which are therefore heavier) to fall out of the air earlier, leaving the finer particles to fall farther from the site of the eruption. One would therefore expect that concentrations of iridium would be greater the farther one went from the site. This is precisely what was found—iridium values increased as the scientists went from 20 to 300 kilometers away from the volcano. The Russian scientists summarized their findings as follows:

> The ash contains 1–4 ppb iridium. Contents increase with decreasing particle size and increasing distance from the source volcano, suggesting that the iridium is deposited by the condensation of volatile compounds on particle surfaces. The authors suggest these findings support a volcanic explanation for the K-T boundary and other worldwide geochemical anomalies.

So, too, did similar findings from the blue ice fields of Antarctica, where an Austrian investigator, Christian Koeberl, looked for iridium concentrations in volcanic dust bands preserved in the ice. Like Greenland, Antarctica is a fine collector of the particulate matter that drifts down from the stratosphere and atmosphere. Each year, particulate matter is buried in that year's snow accumulation and is carried down toward the ocean by the glaciers, eventually appearing as cliffs of blue ice that eventually break off to form icebergs. The exposed portions of the blue ice are tens to hundreds of thousands of years old. The particulate bands within them are presumed in some cases to have originated from volcanoes several thousand kilometers distant.

In the four dust bands that Koeberl sampled, the iridium concentrations were 4 to 8 ppb, with corresponding enhancements of four other trace elements studied—antimony, arsenic, gold, and selenium. The iridium enhancements occurred on particles ranging from 2 to 25 microns. It was another case of high concentrations of iridium on small particles at great distances from the source.

Meanwhile, French investigators Jean-Paul Toutain and Georges Meyer were on the track of iridium concentrations from Piton de la Fournaise, a volcano on Reunion Island. Reunion Island lies southwest of India in the Indian Ocean, atop a mantle plume like that under Hawaii. Its initial eruption at the Earth's surface some 65 million years ago brought about the vast basaltic flow of the Deccan Traps. India at that time was an island chunk of land; the tectonic plate upon which it rests had not yet carried it to its present position as the Asian subcontinent.

The French scientists specifically looked at the iridium concentrations produced by Reunion's recent eruptions that were in the form of *sublimates*—deposits of material formed by the condensation of volcanic vapors. They took three sublimate samples and found iridium values of 4 to 8 ppb, confirming in the process that the iridium was associated with fluorine minerals, and that there were corresponding enrichments of gold and selenium. By way of contrast, they also collected sublimates from volcanic eruptions that were not of hot spot origin; in these sublimates they found no iridium enhancement. They concluded their report: "iridium seems to be preferentially released by hot spot type volcanoes, and its detection at Piton de la Fournaise sublimates provides a positive argument in favour of a volcanic hypothesis to explain K-T boundary events."

From these several findings, it was clear to anyone who wished to see it that certain kinds of volcanoes not only *could* but indeed *had* produced enhanced iridium here and there around the globe—and further, that this process was going on today.

Another question about the Alvarez hypothesis remained unanswered. While meteorite impacts *could* account for enhanced iridium layers here and there, *had* any really done so? Indeed, if meteorite impacts really did enhance iridium layers, wouldn't definite iridium layers be present at meteorite impact sites?

A layer of shattered rock fragments in southern Australia is presumed to be the result of a meteoritic impact. There, sure enough, in 1989, Australian scientists found iridium values of 0.5

to 1.5 ppb. The rock layer was of Precambrian age—that is, some 600 million years old. Given the importance of iridium to the impact thesis, impact proponents searched numerous other meteorite impact sites. But outside the Australian study, *no* iridium enhancements associated with any known impact site have been found.

Little if any of this data was known or suspected when the Alvarez group assumed that the iridium in the Gubbio clays and elsewhere suggested a meteorite impact. But today, and for several years now, the logical assumption would be that a high concentration of iridium found in a geologic section implied the work of volcanoes.

It remains to account for those places around the world where there are high concentrations of iridium from Cretaceous-Tertiary time. If they were of meteoritic impact origin, they should take the form of a *spike*, a nearly instantaneous event in the geological record. On the other hand, if the iridium was of volcanic origin—specifically from the Deccan Trap volcanism that was occurring during the K-T period—they should appear as an *extended deposit* over a few hundred thousand years.

As we have seen, the Deccan Traps were created largely during the magnetic polarity interval 29R, which lasted some 500,000 years. But what were the details of the actual eruption history? Did the eruptions rise to a single peak, or were there several eruption peaks?

Before going any further into this, two factors have to be considered. The first (and it is the lowly sort of phenomenon that a physicist might overlook) can affect the apparent time of an event in the geologic record—that is, worms. Burrowing worms, both on land and in deep ocean sediments, ingest sediment and then egest it. The net effect is that where a worm has burrowed, the sediment is mixed up, a process known as *bioturbation*. In the oceanic regions a worm's burrow today is around ten centimeters deep; compacted over time, a burrow in K-T sediments will be around five centimeters in extent. Given this mixing and sedimentation process, an instantaneous iridium input in the geologic record will appear as a sharp rise, followed in an upward direction by an exponential tail, with a decay constant of the burrowing depth—that is, five centimeters for K-T sediments.

The second factor to consider is that in marine sediments, iridium is found associated with fine clay particles, but not in calcium carbonate planktonic remains. This leads to some tricky problems in analyzing K-T marine sediments. These analyses are usually carried out on bulk samples of sediments, without first removing whatever calcium carbonate may be present. But consider a bulk sample that is 10 percent clay (fine particles) and 90 percent calcium carbonate. An analysis of this sample in bulk will give an overall iridium concentration that is only about *10 percent* of what is contained in the clay fraction.

Particularly in making comparisons among sections, one preferably should report iridium concentrations based on samples that are calcium carbonate–free. A simple procedure exists that avoids any variations that are solely the result of varying amounts of calcium carbonate in the sample: one divides the iridium value by the clay fraction. Thus in a section that is ten parts clay to ninety parts calcium carbonate, one divides the iridium concentration by 0.10. For a sediment that is 95 percent calcium carbonate, one divides by 0.05.

Since the Alvarez discovery in 1980, numerous investigators have made analyses for iridium concentrations at various K-T sites around the world, including those at the original discovery site at Gubbio. The stratigraphy there consists of alternating massively thick layers of limestone (nearly pure calcium carbonate) and far thinner layers (one to two centimeters) of clay. These alternating layers represent several million years of deposition, beginning before and lasting well after the K-T boundary. One of these clay layers corresponds to a foraminiferal K-T change. As mentioned previously, the estimated temporal duration for these clay layers is 10,000 to 20,000 years. In an effort to better define the time duration of the K-T clay layer at Gubbio, the Alvarez group chose to measure its iridium content, on the assumption that all the iridium came from the dilute deposition of meteoritic dust. Then serendipity came into play, much as with Zoller's discovery of enhanced iridium levels in the fine particle emissions from Kilauea. The iridium levels were much higher than expected—and so the impact hypothesis party started.

Eight years later, in 1988, James Crocket of McMaster University in Canada and colleagues went back to Gubbio and sampled all the clay layers extending from 3 meters above to 2 meters below the Alvarez discovery site, as well as the stratigraphic levels 100 to 200 meters down section, for the determination of background values. His results for two geologic sections at Gubbio—Bottacione and Contessa—and quoted on a calcium carbonate-free basis, are shown in the top part of the chart on page 118. The values to the right are background measurements at the Bottacione section. There is a peak for the K-T clay layer of 5 to 15 ppb; the peak itself is broadened out over an interval from about 0.5 meters above to 0.5 meters below the K-T clay layer. In addition, the iridium values extend at elevated levels above background for two meters below the peak and for two meters or more above the peak, with a subsidiary peak at 1.8 meters below the K-T clay layer. Most important, the results at the Contessa section are duplicated by those at the Bottacione section.

The Gubbio section's magnetic polarity interval 29R—which includes the K-T transition—extends over five meters, corresponding to 500,000 years. Thus, the extended distribution of enhanced iridium levels covers 400,000 years or more, and the peak has a duration of 100,000 years.

Understandably, Crocket's findings caused a bit of a stir. They do correspond to the extended deposit that would be expected from the Deccan volcanism; they do not correspond to the spike that would be expected from an asteroid impact. Because of their potential importance in the extinction debate, it was appropriate to go back one more time to Gubbio to repeat the sampling and analyses. Robert Rocchia of the Centre des Faibles Radioactivités in France and colleagues did just that. Their results for the clay partings have been included in the Crocket figure, and they do confirm his results.

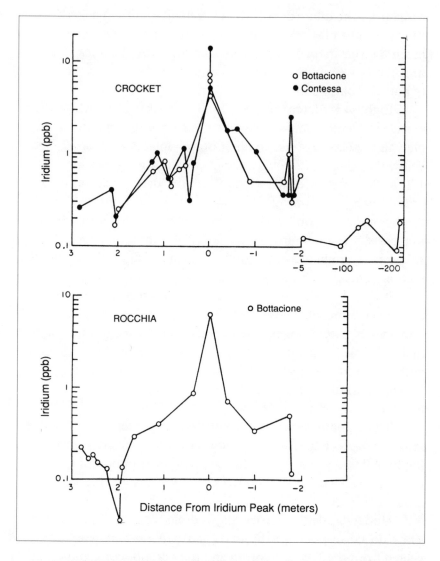

Above: Iridium values for clay layers, quoted on a calcium carbonate–free basis, for two K-T sections in Italy, Bottacione and Contessa. The horizontal scale is centered on the iridium peak. The data points in the panel to the right are background levels farther down the stratigraphic column. *Source:* Crocket et al. 1988.

Below: Iridium values for clay partings, quoted on a calcium carbonate free–basis for the Bottacione section. *Source:* Rocchia et al. 1990.

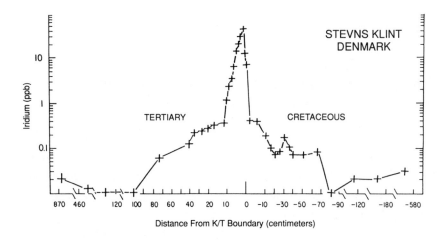

Iridium values, quoted on a calcium carbonate–free basis, for Stevns Klint, Denmark. *Source:* **Rocchia et al. 1987.**

Rocchia went to another site that was included in the original Alvarez study, that at Stevns Klint, Denmark. Here again he found an extended distribution of iridium with a prominent central peak and an indication of a subsidiary peak nearby. From the magnetic stratigraphy the 1.8-meter interval of elevated iridium corresponds to about 100,000 years. For K-T sections at Caravaca, Spain, and Bidart, France, too, Rocchia obtained similar results— an extended iridium distribution with a central peak.

The most detailed study of the iridium variation at a K-T section was carried out in 1985 by Helen Michel of the University of California at Berkeley, from cores at the Deep Sea Drilling Project site 577, which is in the northwestern Pacific Ocean. She was joined in this analysis by the other three original impact theorists, Frank Asaro, Walter Alvarez and Luis Alvarez. (No determinations had been made in these core samples of the calcium carbonate fraction, meaning that the investigators could make no plots on a calcium carbonate–free basis, as Crocket and Rocchia did. Instead, as a proxy, Michel used a plot of iridium normalized to iron—that is, iridium divided by the iron value. Iron was a dominant element in the clay fraction of the cores.) In a cluster of six local cores, she found an iridium peak that occurred within the magnetic polarity interval 29R, during which the sedimentation rate is figured to be 0.76 meters per 100,000 years. This period of time included a 20,000-year interval in which foraminifera changed from mostly Cretaceous types to mostly Tertiary—that is, the K-T transition period.

The iridium peak Michel found, then, occurred over a stratigraphic interval of 3.5 meters, or a period of 460,000 years beginning well before and continuing well after the K-T foraminiferal transition. In addition to the main iridium peak in that period, Michel also found five subsidiary peaks within the main peak: two of these, A and E, she attributed to a possible change in the rate of sedimentation, the others, B, C, and D, perhaps to sampling error. What is important is that peaks B, C, and D could not be explained

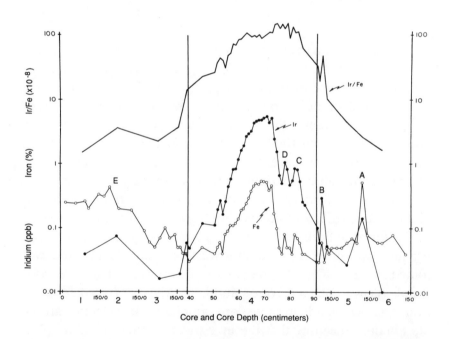

Iridium (Ir) and iron (Fe) values and iridium normalized to iron for Deep Sea Drilling Project site 577, in the northwestern Pacific. Each core has a length of 150 centimeters. Note the expanded horizontal scale for core 4 from 40 to 90 centimeters depth. The K-T foraminiferal transition occurs over the interval from 60 to 75 centimeters in core 4. *Source:* Adapted from Michel et al. 1985.

by either bioturbation (the action of worms) or the iridium spike that would be brought about by a meteorite impact as hypothesized—and that it was necessary to call on a sampling error to explain them.

In fact, what Michel found accorded with what Rocchia and Crocket had found elsewhere: evidence of the deposition of enhanced iridium over an extended period of time—nearly a half-million years surrounding the K-T boundary. Such enhancements, we know, are associated with the fine-particle emissions of present-day volcanic eruptions and can become part of the geologic record. We also know that the Deccan Trap eruptions—eruptions of unprecedented magnitude—occurred at K-T time. It is reasonable, then, to conjecture that the iridium distributions observed by Crocket, Rocchia, *and* Michel are the record of iridium flux from these eruptions. But the Alavarezes and others paid little heed to these findings—even their own.

There are other trace element enhancements at K-T sections in addition to that for iridium. Although it is important to examine them, unfortunately a number of considerations make comparisons with iridium less than satisfactory. There generally are substantial variations in these elements as normalized to iridium— that is, the ratio of element X to iridium from one K-T section to the next, or within a given K-T section. Secondary geochemical processes can favor enhancement of some of the elements over others. Furthermore, for some of the elements only small differences are to be expected from either a mantle or meteoritic source.

Considerably fewer measurements have been taken of the other platinum group elements—ruthenium, rhodium, palladium, osmium, and platinum—at K-T sections, in part because they require more complex separation procedures prior to analytic determinations. The results from the few analyses made are mixed: some suggest an origin from within the mantle, while others suggest a meteoritic origin. For example, two measurements of the ratio of the stable isotopes of osmium, osmium-187 to osmium-186, were made at K-T sections from Stevns Klint, Denmark, and Raton Basin, Colorado. For Stevns Klint, Denmark, the observed value is 1.65, and for the Raton Basin, Colorado, it is 1.29. These bracket a mantle value of 1.44 but are slightly greater than a meteorite value of 1.00.

For several of the other trace elements, the results are equally difficult to interpret. For example, the ratios of zinc to iridium and selenium to iridium are in agreement with the observations from the Kilauea fine particulate matter, but they are too high by a factor of around 50 when compared with meteorites. On the other hand, the ratios of cobalt to iridium and rhenium to iridium are in agreement with meteoritic values, but they are too high by a factor of, again, about 50 when compared with the Kilauea results.

None of these comparisons is definitive, and they may only illustrate our lack of knowledge of what went on in K-T times and during the following 65 million years. But two elemental comparisons stand out like sore thumbs and argue conclusively for a mantle origin. These are the enhanced levels of arsenic and antimony. The accompanying table shows the observed levels of arsenic and antimony and their values normalized to iridium for several K-T sections, as compared with meteoritic values and those from the Kilauea fine particles. The observed arsenic-to-iridium and antimony-to-iridium ratios are too high by a factor of 1,000 over the chondritic meteorite values, but they are in the same ball park as the Kilauea values. That difference cannot be explained away by source variations or secondary chemical effects.

There is an out, however, for the impact proponents. If the hypothesized meteorite landed on an oceanic terrain, which is of basaltic composition and of a mantle origin, it would eject into the atmosphere an amount of material some sixty times that of the impacting body; this material, along with material from the impacting body, would become part of the geologic record. In other words, the arsenic and antimony signals are, indeed, of mantle origin, but from an impact rather than from volcanism. Unfortunately, there is a difficulty with this suggestion. Deformation features in quartz grains, which we cover in the next section of this chapter, are yet another characteristic of K-T sections. Oceanic basalt is devoid of quartz. If the K-T quartz grains are considered to be of impact origin, the impact must have been on a continental rather than an oceanic terrain. The impact proponents can't have it both ways.

Location	As	Sb	Ir	As/Ir	Sb/Ir
Caravaca, Spain	760,000	17,000	57	13,000	300
Gubbio, Italy	18,500	2,460	9	2,100	270
Stevns Klint, Denmark	83,000	9,400	47	1,800	200
DSDP 465A, Atlantic	6,400	670	10	640	67
GPC 3, Pacific	58,000	4,900	10	5,800	490
AVERAGE				4,700	270
Chondrites	1,800	138	514	3.5	0.27
Kilauea	6,000,000	48,000	630	9,500	80

Comparison of observed concentrations of arsenic (As), antimony (Sb), and iridium (Ir) and arsenic and antimony normalized to iridium for various Cretaceous/Tertiary sections, as compared with the corresponding values for chondritic meteorites and the airborne volcanic particulates collected at Kilauea. As, Sb, and Ir values are in parts per billion. From Officer and Drake 1985.

The reader who has come with us this far is to be congratulated. All these fine points about iridium levels and other elements in the K-T sediments are pretty technical. We have tried our best to make it all as palatable as possible, but it is easy to see why newspaper reporters and other journalists writing for laymen have been content to drop the matter. One result, however, is that generally speaking, the iridium layer still stands out in the public mind as an exceptional phenomenon explainable only by means of a meteoritic impact. The Alvarezes' original assertion still reigns popularly as largely unchallenged, in spite of the fact that it does not stand up to any of the actual scientific scrutiny to which it has since been subjected.

Yet another phenomenon, found relatively early in association with K-T sections, was immediately acclaimed as a clear signal of meteoritic impact. It is the planar deformation features (PDFs) in grains of quartz we just mentioned. While this so-called "shocked quartz" received less publicity than iridium, it was billed as valuable evidence of an impact. Again, it is quite a technical matter, not easily explained in a few words or a newspaper paragraph. Subsequent scientific challenges to the value of PDFs as impact evidence have gone largely unreported—even readers of *Science* heard little about them. So they remain, along with the iridium layer, as a kind of mini-mantra, dutifully if only occasionally chanted in whatever media discussions of meteorites and extinctions still take place.

Normally, quartz grains, when viewed under an optical microscope, show clear, smooth surfaces. In some cases, however, one sees single or multiple intersecting sets of lines, like cross-hatching, on the quartz grains. Such grains are found at some K-T sections; they are also found at many other geological settings around the Earth. They are sometimes referred to as *shock lamellae,* but this is a misnomer since not all of them are of shock origin. And here lies the dilemma in trying to interpret them at K-T sections.

Before going on to the K-T observations, let us look at what we do know about these deformation features from laboratory and field observations at known geologic structures.

First of all, in the laboratory, when quartz grains have been subjected to gradually increased pressure up to five to ten kilobars (one kilobar is 14,500 pounds per square inch) over several hours, they have been found to develop both single and multiple sets of lamellae. Approximately the same features show up in rocks that have been deformed at underground nuclear test sites, where the pressure is the same—five to ten kilobars—but is exerted over a period of only seconds or less. They are also formed in laboratory tests of still shorter duration but with far greater pressures—from 50 to 200 kilobars. In other words, these lines and cross-hatching marks occur over a wide range of experimental conditions.

Not surprisingly, they also occur in much the same way in nature: where tectonic plates grind against each other (where pressure is exerted over long periods, deforming the rocks); at explosive volcanic sites where the pressure is short term; and at meteorite impact sites, where the pressure is nearly instantaneous and very great. In the latter—that is, at known impact sites—the lamellae are usually planar (which is to say, straight) and parallel—classic cross-hatching, with the lines two to ten microns apart. These are often referred to as "clean" lamellae, and the accompanying photograph shows examples of them.

The lamellae produced at tectonic and volcanic sites (and at nuclear test sites) are also planar, but sometimes they are slightly to moderately curved. They are also often parallel, but some may show bifurcations and irregular widths of the lines. These so-called "dirty" lamellae can also be seen on page 126. (The findings of PDFs at volcanic sites is of recent origin, the result of studies by Neville Carter of Texas A&M University from Toba volcanic samples, as reported in *Geology* in 1986.)

Yet another kind of quartz deformation occurs (at lower peak pressures than those that cause lamellae) as the crystal structure begins to break down. The grains have a mottled appearance called *mosaicism,* which has been found in both meteoritic and volcanic samples. Indeed, mosaicism is more common than lamellae at volcanic sites—at the high temperatures often associated with volcanism, the crystal structure is annealed, eliminating any traces of lamellar structures.

When it comes to PDFs that were created in K-T times, the picture is further complicated. In North America, PDFs are common, they occur in a thin layer, and they resemble the clean lamellae of a meteoritic impact. Elsewhere around the world, they are rare, occur over a thicker layer (meaning a longer time interval), and resemble the dirty lamellae of volcanic origin. For those trying to decipher its mysteries, nature does not always make things easy.

The first discovery of PDFs at a K-T section was made in Montana in 1984 by Bruce Bohor of the U.S. Geological Survey and colleagues. Similar finds in North America were soon forthcoming. PDFs were common, occurring in about 25 percent of all samples; they occurred in narrow layers about one centimeter thick (and in some cases associated with an iridium peak of one to ten

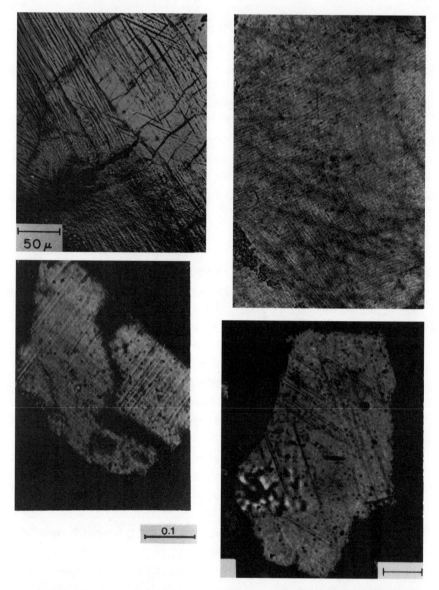

Photomicrographs of quartz grains with planar deformation features. *Upper left:* Sedan underground nuclear test site. Northwest- and west-northwest-trending lamellae. Scale bar 50 microns. *Source:* Bunch 1968. *Upper right:* Tapeats sandstone. Several sets of intersecting lamellae. Length of photograph 650 microns. *Source:* Lyons et al. 1993. *Lower left:* Toba volcanics. Single set of northeast-trending lamellae. Scale bar 100 microns. *Source:* Carter et al. 1986. *Lower right:* Clearwater Lakes impact site. West-northwest- and northwest-trending lamellae. Scale bar 100 microns. *Source:* Carter 1968.

ppb) and were mostly—but not uniformly—planar and parallel (see page 128). All this suggested an impact origin—and after all, impacts do occur.

But a skeptic could point to a nagging uncertainty here. The geology of western North America shows that vast amounts of coal were being deposited in this period. The PDFs and the iridium anomalies all occur in a *tonstein*, a German word meaning "clay stone" and in this case meaning volcanic ash layers that have been altered in the chemical conditions of a coal-creating environment. This means that what appear to be pretty clear signs of a meteoric impact could, instead, be the results of a large volcanic eruption of local origin.

These same PDFs, by the way, also appear to be associated with an extinction—but only a *regional* one, of five kinds of plants out of twenty (as shown in the pollen record). These same five floral plant-types continue to higher stratigraphic levels and younger ages at other K-T sections in North America.

Another feature of these PDFs is that they occurred in rather large quartz grains—some 50 to 60 microns across—and it is hard to imagine such large grains being transported over global distances. So a search began for a regional impact structure. It soon fastened on a circular structure buried underground in Iowa called the Manson structure, which has a diameter of 30 kilometers (which, by the way, is only a fraction of the size hypothesized by Alvarez). In 1989 glass particles associated with the structure were dated at 65 million years ago, and it was promptly heralded as the cause of the North American iridium anomaly and the PDFs. However, subsequent studies in 1993, on sanidine particles from the structure, gave a more reliable age of 74 million years, and it was further noted that the structure itself—with a central core of Precambrian basement rocks that have since been uplifted 6,000 meters—is not your usual sort of impact crater.

Thus did the Manson structure enter and almost immediately exit the K-T stage, leaving us with a yet unknown short-term event of local origin that caused a regional extinction of five plant types—quite possibly a meteorite, but not a very important one.

Photomicrographs of quartz grains with planar deformation features from K-T sections. *Upper left:* Raton Basin, Colorado. Three sets of slightly curved lamellae with bifurcations and variable lamellar widths. Scale bar 50 microns. *Source:* Izett 1990. *Upper right:* Gubbio, Italy. Single set of northeast-trending lamellae. *Source:* Carter et al. 1990. *Lower left:* Chicxulub, Mexico. Multiple intersecting sets of lamellae with slight curvature, bifurcations, and variable lamellar widths. Scale bar 200 microns. *Source:* Quezada Muñeton et al. 1992. *Lower right:* Mimbral, Mexico. Curved lamellae. Length of photograph 270 microns. *Source:* Stinnesbeck et al. 1993.

Elsewhere in the world, a different picture altogether emerges. Neville Carter and colleagues carried out a detailed study of PDFs at the K-T location in Gubbio. PDFs there are rare, occurring in one percent or less of the quartz grains, as is also the case for samples from the Toba volcano. The quartz grains themselves are small, only 10 to 20 microns. Single sets of lamellae are dominant over multiple intersecting sets, another characteristic of the Toba volcanic samples. And the lamellae occur at the peak in the iridium anomaly, as well as over two meters extending both *above* and *below* the peak.

Mosaicism in the Gubbio quartz and feldspar grains is common, and it shows the same stratigraphic distribution as the PDFs and the iridium anomaly—extending from two meters above the iridium peak to two meters below. Something happened there over an interval of around 400,000 years—almost surely something volcanic in origin. Alan Huffman of Exxon Production Research and colleagues found a similar picture at the Deep Sea Drilling Project sites in the southeastern Atlantic.

In short, what really emerged from "shocked quartz" analysis was a relatively murky picture consisting of a probable but not certain impact in western North America and almost certain volcanic origin for PDFs of the K-T era elsewhere on the globe.

NINE
The Silly Season

THE ALVAREZ IMPACT HYPOTHESIS, WITH ITS FOCUS
on dinosaurs, stirred the imagination of even the "veriest dullard"
in its own right. Yet it also led to some astounding secondary
flights of scientific fancy. The speculation and fancy that arose
qualify as prime examples (in Keith Thomson's list of steps) of the
silly season.

Two paleontologists, David Raup and Jack Sepkoski from the
University of Chicago, launched the silly season in 1984 with a
paper they published in the *Proceedings of the National Academy of
Sciences*. They reported on statistical analyses of extinctions among
families of fossil marine vertebrates, invertebrates, and protozoans
over a period of 250 million years. Their results suggested that,
beyond the normal background buzz of species extinction, there
had been twelve peaks or extinction events, with a mean interval
between them of 26 million years. Two of these events, they went

on to say, coincided with the extinctions that had been attributed to meteorite impacts—the Cretaceous-Tertiary and an extinction peak in the Eocene. They concluded their abstract with the statement: "Although the causes of the periodicity are unknown, it is possible that they are related to *extraterrestrial forces* (solar, solar systems, or galactic)."

As a possible explanation for *all* extinction events in the past 250 million years, their paper caused almost as big a stir, and almost as widespread a play in the popular media, as the original Alvarez hypothesis. In the scientific community, some astrophysicists and cosmologists in particular were fascinated and immediately began seeking mechanisms that would bombard the Earth with comets (*not* asteroids as in the original Alvarez hypothesis) on a regular clockwork basis every 26 million years.

As in the short-lived polywater controversy, the game of periodic comet showers was mostly played out in the pages of *Science* and its British counterpart, *Nature*. On April 19, 1984, only two months after the appearance of the Raup and Sepkoski article, *Nature* carried five articles on the subject. Two months might seem an inordinately short period to formulate a sophisticated astronomical theory, but preprints of the Raup-Sepkoski article had been widely circulated long before it actually appeared, giving certain people an inside track.

One triggering mechanism suggested for these cometary catastrophes was an as-yet-unseen companion star to the Sun. This star was presumed to be traveling somewhere outside our solar system in an elliptical orbit. Every 26 million years, it was suggested, this star passes through the Oort cloud, the (hypothetical) home of frozen cometary material outside the solar system. As the star passes through the cloud, it launches the comets on their lethal path. The star, called Nemesis after the Greek goddess of retributive justice, triggered the imagination of the popular press. Nemesis appeared in garish and terrifying full color on the cover of *Time* magazine; and *The New York Times Sunday Magazine* carried an autobiographical sketch by one of the proponents of Nemesis, which was subsequently condensed in *Reader's Digest*, completing the almost instantaneous and nearly total media penetration.

A less dramatic though equally hypothetical comet trigger was assigned to oscillations of the Sun's motion perpendicular to the galactic plane. These oscillations have a periodicity of 30 million years. Yet another trigger was found in the existence of a yet-undetected tenth planet—dubbed Planet X—orbiting the Sun somewhere beyond Pluto. Every 28 million years, Planet X supposedly perturbs some equally undetected disk of comets that lies within our solar system (not in Oort's cloud).

All three of these suggestions, it is important to remember, were triggered not by anything previously imagined, much less suspected, by astronomers as existing somewhere in space, but merely by Raup and Sepkoski's statistical analysis of the marine extinction record. Prior to this analysis, paleontologists had all acknowledged that there were five major extinction events in the past 570 million years. As we have seen, these were the Cretaceous-Tertiary, the Permian-Triassic, the End Ordovician, the End Devonian, and the End Triassic. As shown on page 33 there is no obvious periodicity to these extinctions. They were *recurring* events in the geologic record, but they do not have the same interval between them, which is what is meant by the word *periodic*. Then where did Raup and Sepkoski get their periodicity, not to mention their *twelve* extinction events?

Raup and Sepkoski divided the 250-million-year interval into thirty-nine recognizable stages, giving an average duration for each stage of 6.4 million years. They then plotted the percentage of extinctions for each stage on a time graph, as shown in the figure on page 133. They define an extinction event as one in which the level of extinctions increases from stage X to stage Y and then decreases to the next succeeding stage Z. (Their original graph plotted the percentage of extinctions on a logarithmic scale, but this gives undue emphasis to the lower level extinctions. We have replotted their data on a linear scale, to more clearly indicate relative levels of extinction.)

To start with, we must look at the procedures used by Raup and Sepkoski. They did not include the presently existing taxa—that is, creatures that have survived to the present—when they calculated their extinction percentages. As they pointed out, this

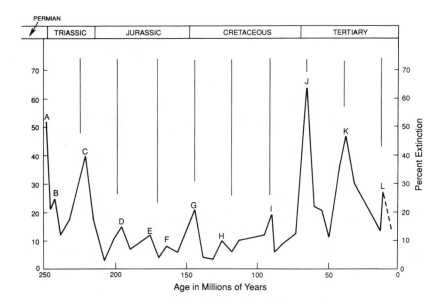

Extinction record for the past 250 million years. The best-fit, 26-million-year cycle, is shown in vertical lines along the top. The relative heights of extinction peaks should not be taken as literal expressions of extinction intensity because the absence of presently existing taxa exaggerates the heights of younger peaks. *Source:* **Adapted from Raup and Sepkoski 1984.**

omission has the effect of exaggerating the heights of the more recent peaks, in particular peaks J, K, and L. For a given stage, for example, if ten families disappeared out of a total of one hundred families and fifty of these families are still present today, their percentage figure would be 20 percent rather than 10 percent if all the families are included. Raup and Sepkoski's plot shows the three major extinction events over the past 250 million years—Permian-Triassic (A), End Triassic (C), and Cretaceous-Tertiary (J). The K-T extinction is not greater than the P-Tr extinction, as shown in the plot. The late Eocene peak (K) is considerably reduced in magnitude. And the middle Miocene peak (L) is questionable, as illustrated by the dashed line in the plot and as pointed out by Raup and Sepkoski as well as others.

The data does show a periodicity of 26 million years. This periodicity is dominated by peaks J, K, and L, which have ages of

65, 39, and 13 million years respectively. However, this is only a *mean*, or average, periodicity; it is not an *exact* periodicity. For example, the interval between peaks A and B is six million years, as compared with 35 million years between peaks H and I. To illustrate the difference between exact and average periodicity: each and every calendar year has exactly 365.25636 days. On the other hand, the first snowfall of winter in a New England town may occur on average on November 15, but that date may vary from sometime in October to sometime in December for any given year.

To continue, there is an inherent difficulty with interpreting any geologic time series. The geologic timescale is actually a bit of a rubber ruler. For the earlier times in the Mesozoic, specific episodes may differ by 5 to 15 million years, depending on whose timescale is used as a reference. Thus the best one can hope to do is to refer to average times between recurring events.

Another difficulty involves determining what significance can be given to the lower-level peaks, specifically peaks D through I. Peaks E, F, and H are close to background levels and may have no great significance. As pointed out by Anthony Hallam of the University of Birmingham in England, the middle Jurassic peak (D) and the late Jurassic peak (G) represent regional extinctions rather than global extinctions—a step up in the process from background, to be sure, but not at the level and global extent of the major peaks A, C, and J.

The most critical commentary on the Raup and Sepkoski analysis came from Antoni Hoffman of the Polish Academy of Sciences as well as, in separate articles, from two other groups. In simple and direct terms, Hoffman's analysis goes along the following lines. Let us presume that the extinction events have a random distribution through geologic time. Then, as we go from stage X to stage Y to stage Z, there are four possible paths, each with the same probability. As shown on page 135, there can be a decrease in extinction percentage from stage X to stage Y and an increase to stage Z; or an increase to stage Y and a continuing increase to stage Z; or a decrease to stage Y and a continuing decrease to stage Z; or an increase to stage Y followed by a decrease to stage Z. The fourth possibility, according to the definition used by Raup and

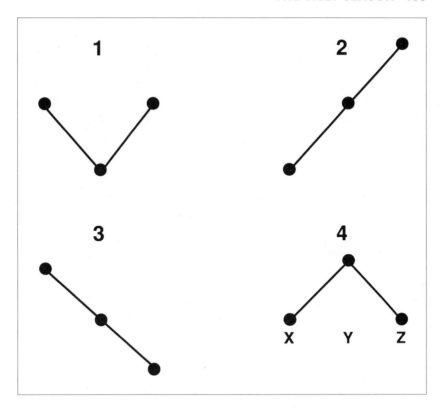

Four possible paths from stage X to stage Y to stage Z for a random distribution. Path 4 represents a mass extinction. _Source:_ Charles Officer.

Sepkoski, would be called an extinction event. With a random distribution there is a one-in-four chance of producing an extinction event. The average length of each stage used by Raup and Sepkoski was 6.4 million years. Thus a pseudoperiodicity is built into any data set of 6.4 x 4 = 25.6 million years, dangerously close to the observed periodicity of 26 million years. In other words, the 26-million-year periodicity could be merely a statistical artifact, and nothing that actually occurred. As Mark Twain quipped in regard to matters of a similar nature:

> Figures often beguile me, particularly when I have the arranging of them myself; in which case the remark attributed to Disraeli would often apply with justice and force: "There are three kinds of lies: lies, damned lies, and statistics."

Beyond the magic of statistics, other reasons to doubt the existence of periodic comet attacks rapidly came to light, in particular a consideration of iridium. After all, if there had been periodic comet showers, there should also be periodic iridium anomalies on Earth with the same 26-million-year periodicity. But quite simply, throughout the Phanerozoic record, there are no other iridium anomalies comparable to that found at the K-T boundary. This latter point was forcefully emphasized when Frank Kyte and John Wasson of the University of California at Los Angeles carried out detailed iridium analyses of a nine-meter section of a core from the central Pacific Ocean. The core was taken from a region of extremely slow sedimentation, and its section extends from 33 to 67

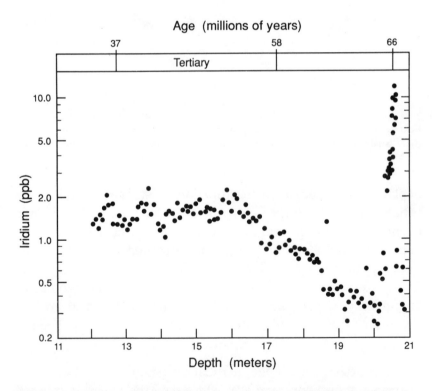

Profile of iridium concentrations in a nine-meter section of piston core GPC-3 from the central Pacific Ocean. The only anomalously high iridium peak is observed at the K-T boundary. *Source:* **Kyte and Wasson 1986.**

million years ago, which includes both the K-T period and Raup and Sepkoski's Eocene extinction event. There is but one iridium peak in the core—the K-T peak—and none corresponding to the Eocene extinction.

Hoffman's critique, which appeared in the June 20, 1985 issue of *Nature*, spelled the beginning of the end for the 26-million-year periodicity notion and the astounding if fragile framework of apocryphal astronomical fate built upon it. It is difficult to imagine some celestial object as grand as a nearby star vanishing altogether without some form of notice, if not mourning, by the astronomical community, but this is exactly what happened to Nemesis. And Planet X vanished from the human imagination even before receiving a proper mythological name.

The editor of *Nature* did take note of Nemesis' passing, however. In an introduction to Hoffman's statistical critique of Raup and Sepkoski, he wrote:

> Those who were last year enthusiastically speculating on
> the mechanism that might have been responsible for mass
> extinctions of living things at intervals of 26 million years
> will be given uncomfortable second thoughts by the article by
> Antoni Hoffman on p. 659 of this issue. For on the face of
> things, Hoffman has undermined the assumption on which all
> the excitement was based, the belief that there is a 26 million
> year periodicity to be explained. But human nature being
> what it is, it seems unlikely that enthusiasts for catastrophism
> will now abandon their quest. . . . The case for Nemesis has
> now wilted.

The silly season continued with two other aberrations from the original asteroid impact hypothesis. One was an attempt to address the problem that it is difficult to explain the observed sequential and extended extinction record with only a single event, wheher it be of extraterrestrial or terrestrial origin. How did the impactors address this problem? Your first guess is correct—they called on six to ten cometary impacts, rather than a single asteroid impact, over a period of three million years. This view was presented in an article in *Nature* by Piet Hut of the Institute for Advanced Study, Walter Alvarez, and others. We quote below the abstract from that article.

If at least some extinctions are caused by impacts, why do they extend over intervals of one to three million years and have a partly stepwise character? The solution may be provided by multiple cometary impacts. Astronomical, geological and paleontological evidence is consistent with a causal connection between comet showers, clusters of impact events and stepwise mass extinctions, but it is too early to tell how pervasive this relationship may be.

The new suggestion did not result from a prediction from the original Alvarez hypothesis. After all, the original hypothesis called on an asteroid impact; the new suggestion calls on a *cometary* impact, a much different proposition. Moreover, the original hypothesis called on a single impact; this suggestion calls on *multiple* impacts at and around K-T times. The new suggestion is simply an attempt to adjust to the observed extinction record.

There are a number of problems with this ad hoc extension. For one, the iridium content of comets, is questionable. For another, the observed K-T iridium record is not in accord with this hypothesis; there simply are no multiple iridium peaks over a three-million-year period. (This particular fact has led to a schism within the impact community: some have stuck with the single asteroid impact, while others switched over to multiple cometary impacts.) For still another, no crater has yet been found for the hypothesized single impact; the problem is only compounded for several impacts. Perhaps the most difficult hurdle is to provide a rational explanation of why the first impact, or impacts, did in only the dinosaurs; the next, the shallow-water shellfish; and the last, the ocean plankton—which is the observed sequence of extinctions.

As mentioned in the previous chapter, geophysical and geochemical data provides another problem for the impactors: the problem of whether the presumed extraterrestrial object hit the land or the ocean. The occurrence of microscopic dynamic deformation features in quartz grains would suggest a continental target. (The basalts that underlie the oceanic regions of the world are devoid of quartz.) But the elemental enhancements of arsenic and antimony would suggest an oceanic target. (Meteoritic material is depleted in these elements, and their presence at K-T boundary sections necessitates an admixture of ocean basalt target material, which is rich in these elements, with the asteroidal impact material.) How has this dilemma been resolved? Once again, your first guess is correct. Some of the impact proponents, notably Walter Alvarez and Frank Asaro, in an article in *Scientific American*, made an ad hoc change in the hypothesis. Instead of one large asteroid impact at K-T time, there were several smaller impacts over a year or so on *both* oceanic and continental terrains. Examples of coeval impacts do indeed exist, such as the ones at East and West Clearwater Lakes in Canada and at Kara and Ust Kara in Siberia; but these are nearby impact sites, not sites over distances between continents and ocean basins.

Both of these suggestions—multiple impacts over a three-million-year period and multiple impacts over a year or so at K-T time—were fabricated to accommodate known facts. Neither was the result of a prediction from the original hypothesis.

A corollary to the Alvarez hypothesis emerged quite early, based on the finding of minute amounts (in parts per million) of carbon black, the carbon left after burning, at three K-T sections of marine origin. This carbon, it was concluded immediately, had to be the result of huge, wide-scale burning of the world's forests. Since green and growing trees do not burn very well, the suggestion was made that the world's "forests were killed and freeze-dried by the initial darkness and cold following the impact, and were then ignited by lightning after the skies had cleared and thunderstorms had resumed."

The global wildfire scenario was duly reported as a new discovery and confirmed fact to the public by some of the science media. In the October 1985 issue of *New Scientist* we read:

Huge fires like those that would follow a nuclear war may have wiped out the dinosaurs. The fires were started by the impact of a meteorite hitting the Earth, a collision many scientists now believe occurred about 65 million years ago (Mya).

Three researchers from the University of Chicago have new evidence for such a fiery apocalypse—a type and level of soot in ancient sediments that could have been produced only by massive fires.

And in the October 1985 issue of *Science News*, a popular science weekly, similar rhetoric matched the supposed apocalypse:

Just when it was looking really bad for the dinosaurs, it got worse. Another element has been added to the already dire scene painted of the world 65 million years ago. Some scientists posit that is when an asteroid or a torrent of comets pelted the planet, wiping out the dinosaurs and hordes of other life (*SN*: 6/29/79, p. 356). In addition to the possible dust clouds, blast waves, tidal waves and poisonous gases triggered by the impact, researchers at the University of Chicago have added yet another deadly plague: continent-sized wildfires churning out massive clouds of soot that engulfed the globe.

Their findings not only enhance understanding of the forces that drove the dinosaurs to extinction, but they provide a much needed quantitative basis for studies of future cataclysms that could befall the earth.

Just how any event, impact or otherwise, that took place in, say, China could kill and dry out the trees in New England was never explained. But beyond that commonsense problem, this notion had other problems. In the first place, the original carbon findings were from marine sediments. Was there any carbon evidence from terrestrial sediments, which is after all where the forests were? Presumably the burning of all tropical and temperate forests would have left a great deal of carbon (not to mention charred dinosaur bones) lying around. But in fact no such carbon layers were found in K-T sediments in North America or other terrestrial sites.

Furthermore, a significant failure of reasoning was associated with the original data on which the global forest fires hypothesis was built. The original data had come from K-T sections taken in Denmark, New Zealand, and Spain. Only at the Danish site, however, were background values of carbon also taken. These values ranged from 1,040 parts per million (ppm) to 950, while the K-T value was 3,600—an anomaly of 3.6 over a background of one. To understand the significance of an anomaly requires knowledge of sedimentation *rates*. For the sake of argument, let us suppose that sedimentation rates slowed down for some reason at K-T time by a factor of ten. Then the concentration of carbon black in the K-T sediments would be ten times higher.

In the Danish section, the K-T boundary is marked by a clay layer known as the Fish Clay, since it contains a lot of fossil fish remains. Above and below the Fish Clay are more normal calcium carbonate types. The Fish Clay represents a condensed section that had a sedimentation rate seven times less than the carbonate sections above and below it—in other words, it was deposited at a significantly slower rate. If the concentration of carbon black in that section—3.6 to 1—is then multiplied by the sedimentation rate of 1 to 7, the carbon black anomaly not only disappears but, if anything, it becomes a *negative* anomaly.

So much for the global wildfire apocalypse, and the "quantitative basis" it provided "for studies of future cataclysms that could befall the earth."

Another finding from the Danish K-T section at Stevns Klint was less sensational, perhaps, and remains a bit ambiguous. Investigators from the Scripps Institution of Oceanography found two unusual organic compounds at the site—alpha-amino-isobutyric acid and racemic isovaline. These are amino acids that are rare in Earth materials but are apparently common in meteoritic material.

No one has been able to explain how any organic compounds, such as these amino acids, could have survived an impact; other equally fragile or volatile compounds and elements have been sought but not found. The high temperatures and strongly oxidizing conditions in a fireball argue against their survival. More likely, they would have been incinerated or atomized. Beyond that, the amino acids have an unusual stratigraphic distribution. In the Danish section they are abundant above and below the K-T boundary but are completely absent in the Fish Clay itself—another *anti-anomaly*. Perhaps they diffused upward and downward after the supposed impact, but such diffusion is as difficult to accept for this phenomenon as it is for iridium and other elemental enhancements. So their distribution in the sediments remains an open question, and their origin, presently unknown, may come to light in later findings. The progenitors for these acids are known to be formed in coal gasification, for example, when coal is converted to methane and other cleaner burning fuels. Perhaps the amino acids at Stevns Klint were formed by a volcanic or other magmatic intrusion into a coal field.

Any impactors who were alarmed by the overwhelming accumulation of evidence for intense volcanism at K-T time may have been reassured by a wondrous hypothesis that appeared in the silly season. It represented a best-of-all-worlds scenario, with something to satisfy everyone. It went something like this: The K-T impact triggered the Deccan volcanism. It was the combined effects of the impact and the volcanism that were the killing mechanisms for the mass extinctions. Ergo, everybody is happy—volcano and impact enthusiasts alike.

There are a number of problems with this best-of-all-worlds scenario, however. For one, the mantle plume that first erupted at the Earth's surface some 65 million years ago and formed the Deccan Traps was located at Reunion Island in the western Indian Ocean. No impact crater is associated with this site. Strike one.

An impact of the size presumed would have formed a crater some 200 kilometers in diameter and 15 kilometers deep—making a major hole in the Earth's surface. But no one, not even with the help of computers, has been able to envisage what dynamics would have triggered a plume of lava to ascend from the core-mantle boundary, which is located some 2,900 kilometers below the surface, halfway to the center of the Earth. Strike two.

Even if these difficulties could be overcome, a third and implacable one remains. The peak of the iridium anomaly and the time of the presumed impact are located about one-third of the way down in the magnetic polarity interval 29R. But the Deccan Trap eruptions began in magnetic polarity interval 30N, several hundred thousand years earlier. In other words, the Deccan volcanism started *prior to* the K-T event. No K-T event could have been the triggering mechanism for the volcanism. Strike three.

On the other hand, in a silly season, who wants to listen to the umpire?

TEN

The Missing Crater

FOR ALL THE CASCADING DIFFICULTIES THAT geologists and others have presented for the Alvarez hypothesis, the impactors would have *something* of a case if they could point to a massive impact crater, preferably several, dating to the proper time in the geological record. They have searched far and wide around the world for evidence of even one such crater, but sadly for them, they have come up wanting.

This failure has not been widely noted in the press; instead, most popular accounts still report that the K-T crater has been located on Mexico's Yucatán Peninsula in a circular feature 200 kilometers in diameter called the Chicxulub structure. Before Chicxulub, several other candidates around the globe were proposed: at the Gulf of St. Lawrence, Hudson Bay, Iceland, and the Andaman Basin in the Indian Ocean, as well as the aforementioned Manson structure in Iowa. Each has been rejected for sound geological reasons and abandoned by the impactors.

The most recent crater hunt concentrated on the Caribbean and Gulf of Mexico region. Various geologic sections—from

Cuba, Haiti, Deep Sea Drilling Project sites 536 and 540 in the Gulf of Mexico, and Mimbral in northeastern Mexico—were reinterpreted to represent deposits from a nearby impact. These sectons led to a frantic search for a nearby circular structure 200 kilometers in diameter that might be identified as the K-T impact site. Four sites were suggested—Isle of Pines off Cuba, Massif de la Hotte on the southern peninsula of Haiti, the Colombian Basin in the western part of the Caribbean Sea, and Chicxulub.

The whole escapade—the *Caribbean caper,* as some have called it—was played out over a relatively short period of time, from 1990 to 1994. Most of the articles that favored an impact in the Caribbean appeared in either *Science* or *Nature;* all of the rejoinders appeared in the more conventional science journals.

In 1990 *Nature* published an article suggesting that an extensive bed of boulders in the western portion of Cuba was an "impact ejecta blanket." A large crater was presumed to lie southwest of Cuba, at the Isle of Pines. No such "impact ejecta blanket" had ever been reported before, so the authors were at some liberty to imagine what such a thing looked like.

One of the authors of this article was not a scientist at all but a former fellow at the Kennedy Center for International Relations. We mention this not to be snobby: some major advances in geology, such as continental drift, have been made by individuals outside the discipline, as in other sciences. It merely highlights how widespread was the interest in the dinosaur extinction debate. A truly problematical matter needs mentioning with regard to the article in *Nature,* however: neither of its authors had visited the site of this boulder bed. All their information came from the published literature. Geologists *do* tend to get snobby about such omissions, believing that field geology really must be done in the field. They hark back to a statement made by the great nineteenth-century Swiss-American geologist and naturalist, Louis Agassiz: "Study nature, not books."

In any event, Manuel Iturralde-Vinent of the Museo de Historia Natural in Havana *did* subsequently go into the field to look at this section, appropriately known as the Big Boulder Bed. It is a hard calcareous sandstone of Maastrichtian age that contains large boulders up to a meter in diameter. But, as Iturralde-Vinent found, the boulders are of a local origin, the product of *in situ* weathering—an exfoliation process still going on *today* and not inherited from any extraordinary event at the K-T boundary. Other investigators confirmed his findings.

The second geologic section to receive renewed attention was near the town of Beloc on the southern peninsula of Haiti. This candidate was of far greater interest than the Big Boulder Bed. Here early investigators had found brown globules, about a millimeter in diameter, with a palagonite composition. Palagonite is formed when volcanic glass combines chemically with water as it seeps into the glass via its cracks and vesicles. The final decomposition of volcanic glasses is a kind of clay called smectite clay. In addition, at the Haiti site, there are a few black glass spherules and some rare yellow glass spherules that have not been altered by any activity since they were formed.

Now, earlier investigators had already interpreted these materials as being of volcanic origin. But in 1990 the brown globules were reinterpreted by Alan Hildebrand and William Boynton of the University of Arizona as being "clay altered tektites." Tektites, as we noted earlier, are odd-shaped bits of glass often associated with meteorite impacts, and so for a brief moment this site was taken to be meteoritic in origin. But further separate studies— by John Lyons of Dartmouth College and Celestine Jéhanno and colleagues from the Centre des Faibles Radioactivités—belied this interpretation.

The problem came down to the brown globules and the black and yellow glass spherules. Volcanic glasses are quite common; impact glasses (tektites and microtektites) are also quite common and have been found in strewn fields at four locations around the world, with ages from 0.7 to 35 million years. Volcanic glasses and tektites are significantly different affairs, however, and the third wave of interpretations at Haiti was unambiguous.

The palagonite alterations of the brown globules (amounting to 90 percent of the deposit) are characteristic of volcanic glass. Tektites, by contrast, are resistant to any alteration, and no altered tektites have ever been found. Second, some of the globules were broken, and all the broken ones were hollow. But no known tektites are hollow. Third, the black spherules—amounting to some one to five percent of the entire deposit—are of a particular composition known as andesitic-dacitic, which is a known product of volcanoes. Tektites, however, have a higher silica content than the black spherules. Fourth, the glass spherules contain vesicles—another characteristic of volcanic glasses—while tektites are characteristically devoid of vesicles and bubbles.

Furthermore, other features of the Haiti glasses—their iron reduction state, the lack of a high-temperature form of fused quartz called lechatelierite, and the presence of sulfur in the yellow glasses—are all inconsistent with the formation of tektites.

Still further, the very nature of the stratigraphy argues against a single event producing the gobules, whether an impact or any other event. At the nine outcrops studied, the sections have a nominal but variable thickness of 30 centimeters; within each section the brown globules may occur as *two to six* graded deposits, some of which are separated from one another by several centimeters of bioturbated limestone. Some of the graded deposits contain black glass spherules as well as brown globules; others do not.

To interpret the Haiti deposit as of impact origin, one has to assume an abundant alteration of tektites that has never been observed elsewhere; tektites of a composition and vesicularity never before observed; and a formation process at lower temperatures than have ever been associated with tektites. In short, it requires inventing an entire new geology to fit the observations. Just such an invention was the "clay altered tektites" interpretation. There is, in fact, no such thing.

The next geologic sections to be restudied were from Deep Sea Drilling Project sites 536 and 540 in the Gulf of Mexico. Both sites are adjacent to the Yucatán Peninsula and the Chicxulub structure, the location that in 1995 was still favored as the K-T impact site. If a large impact had occurred on the Yucatán Peninsula, impact debris would have been spread throughout the Gulf of Mexico and the Caribbean, particularly at nearby locations like sites 536 and 540. The "geology" of these sites followed much the same script as for the Haiti site—a play in three acts: an original interpretation, a revisionist interpretation, and one, as it were, by a subsequent truth squad.

At both these drilling sites there was a 2.5 meter-thick sequence of crudely graded deposits from around K-T times. Both deposits contained limestone fragments, altered volcanic particles now of a smectite clay composition, and volcanic glass shards. Both, upon original inspection, were said to be of volcanic origin. But in 1992 Walter Alvarez and some colleagues asserted that the 2.5-meter sequences at both sites represented "ejecta from a nearby impact crater, reworked on the deep floor by the resulting tsunami." As at Haiti, the smectized clay was reinterpreted as "clay altered tektites," and the scientists reported grains of quartz with planar deformation features. In addition, they found a small iridium enhancement of 0.6 ppb, with a narrow peak in a clay layer above the sequence in question. They interpreted it as evidence of an impact.

Then in 1993 Gerta Keller and her colleagues took another look. They confirmed the presence of smectized clay and other volcanic products, including volcanic glass shards, but they found no tektites or microtektites. The quartz grains had multiple but irregularly spaced grains of fluid inclusions, but none had deformation lamellae. The investigators did confirm the tiny iridium peak, but an enhancement that small is hard to interpret; it could be a reduced K-T anomaly, but it also could be the result of sedimentation rate or subsequent geological changes.

At the same time, Keller and colleagues did paleontological studies of closely spaced samples from the cores from both sites. They actually found a *hiatus* in the sedimentation at K-T time, which is also found at fourteen nearby Deep Sea Drilling Project sites. Sedimentation was therefore not continuous across the K-T boundary here. Furthermore, the 2.5-meter sequence that Alvarez studied was of *late Maastrichtian* rather than K-T age. It happened *before* the K-T transition.

It is not uncommon to encounter such hiatuses in the paleontologic record; they can result from periods of nondepositon or of subsequent erosion of deposited material by bottom currents. But before making detailed studies at a given site, it is essential to determine from the site's paleontology that one indeed has a record at all. No geologic record, no data. Quite simply, if there is a gap in the sedimentary record for a given geologic section, you can't make any pronouncements about what might have happened then. So much, then, for sites 536 and 540.

In the Arroyo el Mimbral, in northeastern Mexico, there is an outcrop along a cliff face that in the 1930s John Muir, a petroleum geologist, described as a channel deposit of approximate K-T age. The deposit extends for some 150 meters along the outcrop at thicknesses ranging from five to eight meters. It represents a former channel that was later filled in with clastic materials—that is, fragments of former rocks. The deposit consists of three distinct layers—a spherule-rich layer at the bottom, a laminated sandstone layer in the middle, and a rippled sand and shale layer at the top; nothing particularly extraordinary.

But in 1992, Jan Smit of the Free University of Amsterdam and his colleagues published an article in *Geology* that interpreted all three layers in the channel to be of impact origin. The bottom layer, the one with the spherules, was said to be "a channelized deposit of proximal ejecta" from a nearby presumed impact site. The middle layer —the laminated sandstone—became the result of "megawave backwash that carried coarse debris from shallow parts of the continental margin into deeper water." That is, it was a result of a tsunami caused by an impact in the ocean. And finally the top layer was seen as "deposits of oscillating currents, perhaps a seiche"—meaning the subsequent oscillations of a tsunami.

Once again, a whole new geology had been invented.

But then Wolfgang Stinnesbeck of the Universidad Autónoma de Nuevo Leon in Mexico led a group of colleagues from Mexico, the United States, and France back to the site, and they published their results in *Geology* in 1993. In the lower layer they found one rare volcanic glass shard for every 200 grams of bulk samples, along with abundant spherules. The spherules were generally one to five millimeters in diameter and were commonly filled with calcite, though some contained clasts of limestone and, in a few cases, foraminifera. They found no tektites. (There is, by the way, no such thing as a tektite with foraminifera inside it.) So these phenomena had to be of local origin, not something thrown out of an impact crater.

They also found quartz grains in the lower layer, a few of which did have PDFs. But these PDFs were curved and of variable widths, as shown on page 128 (lower right)—the kind that are of tectonic/volcanic origin. The same PDFs were found in the same rare quantities in samples from two to three meters *above* and *below* the channel deposit.

An iridium enhancement of 0.8 ppb was found in a clay layer above the channel deposit, but no such anomaly occurred in the channel deposit itself. As at sites 536 and 540, the enhancement was in a thin clay layer, perhaps the result of reduced sedimentation or some K-T event, but hardly what an impactor would like to think arises from a nearby impact.

Each of the three layers at this site was characterized by a distinct lithology, or rock structure, and a distinct mineralogy. But if they all resulted from an impact and subsequent tsunami, all three layers would have been of the same composition. Furthermore, there are discrete intervals of bioturbation within the sequence, and on an exposed ocean bottom, such worm burrowing takes time. Also, within the spherule-rich bed at the bottom there is a 20-to-25-centimeter-thick limestone layer. Such a layer has no business in an impact scenario, and neither do the obvious erosional discontinuities between the three layers. In other words, in geologic terms, the three layers were not formed by an instantaneous event. And if that weren't enough, upon careful analysis the entire sequence turned out to be of latest Maastrichtian age, with the foraminiferal K-T boundary occurring just above the channel deposit.

In summary, there was no evidence of a nearby impact event from these four geologic sections.

Then what of the four suggested impact sites? The Isle of Pines was suggested as a result of the interpretation of the Big Boulder Bed as an "impact ejecta blanket." There is, indeed, a curved fault system on the island, but it is of Tertiary rather than K-T age. The Massif de la Hotte suggestion came as a result of the interpretation of the Haiti section as "clay altered tektites." There is no circular structure of the size required and of K-T age on the southern peninsula of Haiti.

The Colombian Basin in the western Caribbean was also suggested as a possible impact site. The basement topography there has a semicircular aspect of about 300 kilometers in diameter, and it was suggested that this topography was formed by an impact at K-T time. Unfortunately, from the Deep Sea Drilling Project results and continuous seismic profiling between sites and across the Colombian Basin, we know that there are *flat-lying* sediments of Upper Cretaceous age overlying the basement. Thus it is a stratigraphic impossibility for the topography of the underlying basement itself to have been formed by a *later event* of K-T age. Three down, one to go.

The Chicxulub structure was first suggested as a possible impact site in 1981, at an oral presentation at the annual meeting of the Society of Exploration Geophysicists. That suggestion received little attention until the 1990s, when it became the *preferred* site for the hypothesized impact. It therefore deserves special attention, and a detailed description of what is known about the structure— perhaps more than the reader might wish. But bear with us, for this is the end of the line.

The Chicxulub structure was first observed in outline from gravity and magnetic surveys. Gravity surveys measure small changes in the Earth's gravitational attraction; a positive gravity anomaly indicates that there is a body of greater density beneath the gravity station than in surrounding areas. Magnetic surveys measure changes in the Earth's magnetic field; a positive magnetic anomaly indicates a body beneath the measuring station that contains a greater amount of magnetic minerals, usually magnetite. Such surveys, in and of themselves, do nothing more. They have long been a standard geophysical tool in the exploration for petroleum and mineral ore reserves; the Chicxulub surveys were conducted for Pemex, the national petroleum company of Mexico. But gravity and magnetic surveys are by no means categorical discovery methods for impact structures. (In fact, there are other circular magnetic and gravity anomalies of comparable magnitude and extent on the Yucatán Peninsula, in particular the Tizimin gravity anomaly 150 kilometers east of Mérida and the Puerto Juárez magnetic anomaly on the northeastern coast.)

Exploratory drilling in the 1960s by Pemex showed that the geophysical anomalies at Chicxulub are related to an andesitic, i.e., volcanic, body at a depth 1,200 to 2,000 meters. As shown in the figure on page 153, well data across this part of the Yucatán Peninsula shows a continuous sequence of Tertiary and Cretaceous limestones and dolomites. The Cretaceous-Tertiary boundary is the line between Eocene-Paleocene and Maastrichtian in the diagram. The Yucatán No. 6, Chicxulub No. 1, and Sacapuc No. 1 wells on the Chicxulub structure have Upper Cretaceous sediments *overlying* the andesite, and the Yucatán No. 6 well penetrated the andesite and bottomed in Cretaceous limestone and anhydrite. (Anhydrite is an evaporite deposit consisting of calcium sulfate.)

Now, the data in this figure alone is sufficient to rule out the Chicxulub structure as an impact site of K-T age. Had there been a K-T impact at Chicxulub, all of the overlying Upper Cretaceous sediments would have been blasted out, and the anhydrite at the bottom of the Yucatán No. 6 well would have been vaporized. For an impact that created a crater 200 kilometers in diameter, the depth of the crater would be *ten kilometers or greater*—far in excess of the depths shown in the figure. The crater would be gigantic, and the infilling debris would consist primarily of breccia from the underlying basement.

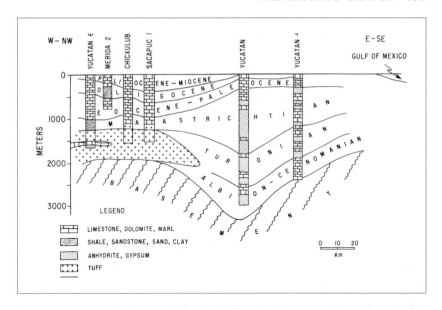

Inferred correlations from well control data along an east-west section in the northern part of the Yucatán Peninsula. The Chicxulub structure is to the left. *Source:* **Lopez-Ramos 1975.**

More on the Chicxulub structure: The late Arthur Meyerhoff was a consulting geologist to Pemex at the time the wells were drilled and was closely involved in the biostratigraphic correlation of the Yucatán wells. Fortunately, he retained in his files the well log and core descriptions from the critical well, Yucatán No. 6, drilled in 1966. This data was published in the January 1994 issue of *Geology*, and the part of the well log covering the volcanic sequence is repeated here as the figure on page 154. (Pemex itself has been reluctant to share its file data with the general scientific community.)

With reference to this figure, the lithologic descriptions as given on the original well log by José Maldonado are as follows: A—Upper Cretaceous (Campanian) limestone; B—emerald-green bentonitic (altered volcanic ash) breccia; C—fine-to-medium-grained, gray-green sandstone containing thin intercalations of light brown to buff and gray bentonite; D—siliceous bentonitic breccia of light-green, dark gray to black, and reddish-brown color, including small-to-medium-sized and a few large fragments

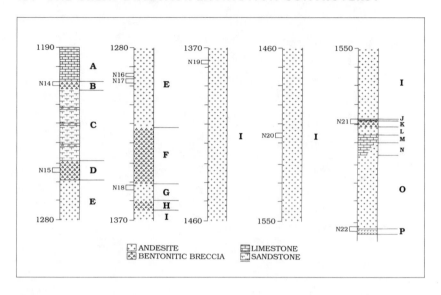

Yucatán No. 6 well log. The well bottomed at 1,645 meters. N14 to N22 mark the depths at which cores were taken. *Source:* **Meyerhoff et al. 1994.**

of limestone; E—andesite; F—bentonitic breccia; G—andesite; H—bentonitic breccia; I—igneous rock; J—limestone; K—volcanic breccia; L—andesite; M—white crystalline limestone, very fine, saccharoidal; N—cream white limestone, very finely crystalline with traces of andesite; O—andesite; and P—anhydrite. The log shows a sequence of at least six major volcanic events—unit O, units L and K, units I and H, units G and F, units E and D, and unit B. Within the sandstone of unit C there are also three bentonitic intercalations.

The critical cores within the andesitic interval are cores N14 through N22. Their descriptions at the time they were recovered are as follows: N14 and N15, bentonitic breccia of andesitic or basaltic composition; N16 through N20, basaltic andesite or andesitic basalt; N21, *interbedded* limestone with late Cretaceous microfossils, volcanic breccias, and andesite; and N22, mainly anhydrite and limestone of Turonian age. This volcanic sequence extends from *Campanian to Turonian in age, that is, from 80 to 90 million years ago.*

There is *no* similarity between this well log and the simple succession encountered in impact melt sheets. Impact melt sheets are formed after excavation by a meteor impact and are formed by very high temperatures and very rapid melting of the target basement rocks. For some larger craters the melt sheets can reach appreciable thicknesses—Manicouagan, 100 to 200 meters; West Clearwater Lake, 120 meters; and Mistasin, 80 meters. (These measurements refer to a *single* melt sheet and not *six* volcanic layers.) One of the outstanding characteristics of melt sheets is that they are *chemically homogeneous* and represent the aggregate composition of the target country rock. Chemical analyses of cores from Chicxulub show a wide range in chemical composition; they do not show the chemical homogeneity characteristic of impact melt sheets.

Rare sets of single and multiple intersecting lamellae were found in in quartz grains from these cores. As shown in the figure on page 128 (lower left), the lamellae are curved and have variable widths. They are of the volcanic/tectonic type.

There have also been later *thermal* events in the Chicxulub volcanic sequence—which can, and apparently have, led to a resetting of the potassium-argon clock. Plagioclase (a common volcanic mineral) has been replaced by albite (a complementary, secondary mineral). Radioactive age determinations from glass samples give values ranging from 58.2 to 65.6 million years—ages much younger than the 80 to 90 million years from the paleontologic studies but in accord with what would be expected from samples that had been thermally altered with loss of the argon content. The same kind of age discordance occurred from the Manson structure in Iowa, once considered a K-T impact site, where recent age determinations of 74 million years from sanidine replaced previous glass ages of 65 million years.

In summary: Either the oil company geologists and paleontologists could not tell the difference between a limestone, sandstone, breccia, crystalline igneous rock, and anhydrite and could not distinguish between Turonian, Campanian, Maastrichtian, and Paleocene microfauna—or the Chicxulub structure is a *volcanic sequence of late Cretaceous age*. It is not an impact melt sheet of Cretaceous-Tertiary age. We conclude with a news brief from the British publication *Geology Today*.

The non-excavating impact: Here's quite a bomb—until it gets defused, that is. It's probably true to say that, apart from a bunch of reactionary paleontologists (only joking), most Earth scientists have come to accept that an asteroid impact directly or indirectly did for the dinosaurs and other species 65 million years ago, if only because they've been beaten into submission by the endless barrage of propaganda in its favour. And the word that's been in their ears constantly for the past few years is "Chicxulub"—the site of the 200-km-diameter, 65-million-year-old structure (on the Yucatán Peninsula in the Gulf of Mexico) thought to have been produced by the impact. But wait; hear the other side first. Meyerhoff, Lyons and Officer (*Geology,* v. 22, p. 3, 1994) draw attention to a previously unpublished well log for the Yucatán No. 6 well, drilled in the centre of the Chicxulub crater in 1966 by the oil company Pemex. The well penetrated an orderly sequence of Pliocene, Miocene, Oligocene, Eocene and Paleocene sediments and then 350 m of uppermost Cretaceous sediments with Maastrichtian microfauna above and middle Campanian fauna below, all overlying an extensive volcanic sequence of andesite (mainly) interlayered with chiefly bentonitic breccia but also a layer containing Late Cretaceous fossils. You see the problem, surely. The Tertiary sediments, being younger than 65 million years, could be there whether or not there had been an impact; but the strike of an asteroid large enough to produce a crater 200 km across would have excavated at least 10 km of the upper crust, destroying the uppermost Cretaceous layers. They are still there; ergo, no impact. Moreover, there is absolutely no sign of any impact melt sheet, the chemically homogeneous thermal product of an impact, which for large impacts can be quite thick (e.g. 100–200 m in the Manicouagan crater). If the Chicxulub structure turns out not to be an impact crater after all, it won't kill the K/T asteroid-impact hypothesis, but it will make quite a few asteroid addicts look naively, er, overenthusiastic.

During K-T times a great deal was happening to the Earth that affected the overall conditions for life on the planet, as well as creating regional and local upheavals. But so far as the geologic record gives us a window on those times, one of the things that did not happen at the K-T boundary was an impact by a gigantic meteorite. Nor is any such deus ex machina necessary to explain the mass extinctions that took place around K-T times—or any other extinctions.

In fact, the Earth did it.

ELEVEN
What Did Happen

MUCH OF THE DEBATE THAT HAS TAKEN PLACE
since the Alvarez hypothesis was announced in 1980 has focused
on such matters as iridium and planar deformation features in
quartz grains. Both are important, but they are *minor* constituents of the geologic environment. The iridium anomaly, after all,
is measured in parts per *billion,* and except in western North
America, PDFs in quartz grains are quite rare. On the other
hand, *major* events with *major* effects took place at the K-T
boundary—as they have before and since. Some of these events
were of a massive but regional scale, yet others were of an even
greater, global scale. They have all been noted in passing
throughout our discussion so far, but now it is time to present
them in a single sequential account—one that, among other
things, provides a perfectly rational explanation for the extinctions at issue in this book.

The Earth works like this:

The Earth's outer surface, or crust, is divided into a number of plates. These plates move horizontally at rates of a fraction of an inch to a few inches per year. New plate material is formed at their originating ends, and old plate material is shoved back into the Earth at their trailing ends. The new plate material consists of molten magma brought up from the depths, particularly along midocean volcanic ridge systems. The old material returns to the Earth's mantle, principally along the major earthquake zones and deep sea trenches surrounding the Pacific. The plates themselves move as rigid slabs over a viscous underlying mantle, driven (it is widely believed) by thermal convection currents in the upper mantle (see figure below). This is the essential picture drawn by plate

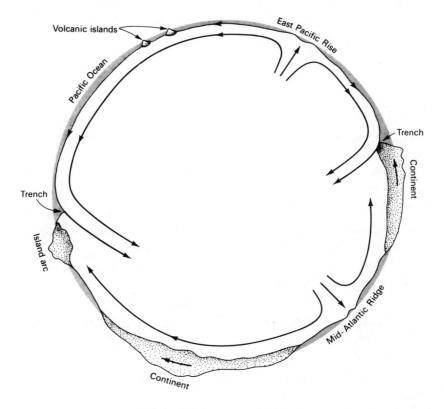

Schematic cross-section of the Earth, based on the seafloor spreading hypothesis. *Source:* **Uyeda 1971.**

tectonic theory, which is universally accepted among today's scientific communities.

Volcanism is directly associated with the midocean ridges, with molten material filling the gaps that occur as the seafloor spreads and two plates move off in opposite directions. The ridges are thought to be fed, either directly or indirectly, with molten material that comes up as giant plumes from a great depth. Called *mantle plumes,* they originate near the boundary between the lower, solidified mantle and the liquid core of the Earth—a boundary that exists about halfway from the Earth's surface to its center. The molten material in the plumes is basic (as opposed to acidic) in composition, dominated by heavy metals, and enriched in sulfur dioxide—which, along with the contained carbon dioxide, chlorine, and water, is vaporized when the molten magma erupts at the Earth's surface. These latter components of the magma are called its *volatile constituents.*

When two plates move away from each other, each crashes against at least one other plate at its leading edge—a *subduction zone,* where one or both plates plunge back into the upper mantle. When a descending plate reaches a sufficient depth, its temperature gets high enough to bring about at least partial melting, which in turn produces chambers of magma that tend to rise up. This rising magma, in turn, can produce volcanism at the surface. Volcanoes in subduction zones (such as Indonesia) typically spew forth debris that is acidic and lighter, the products of the subsumed and overlying crustal materials. One such group of volcanoes is that around the Pacific Ocean, known as the Ring of Fire.

What, more precisely, is the driving force within this plastic layer of the upper mantle that causes the plates to move around?

Several mechanisms have been suggested, and one, *gravitational circulation,* seems to give a reasonable explanation. Fortunately, it does come up with the correct order of magnitude for the plate velocities.

Gravitational circulation is a common phenomenon and can easily be seen in a typical river estuary. Here riverine freshwater of one density meets with the saline sea water of another density. The difference in density is very small—a matter of some two-hundredths

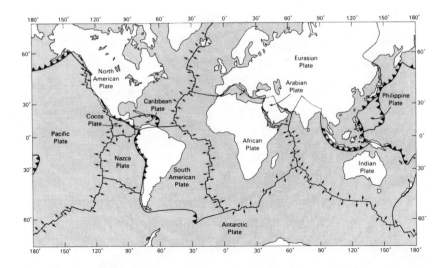

Twelve tectonic plates and their motion. Triangles along boundaries indicate direction of underthrusting, where a downgoing slab can be identified by the occurrence of intermediate or deep-focus earthquakes. Small arrows on ridge boundaries indicate approximate direction of relative motion. *Source:* **Uyeda 1971.**

of a gram per cubic centimeter of water—but it is enough to create a *longitudinal density gradient flow:* the fresh (river) water flows out on top, and the saline (ocean) water flows in on the bottom.

Mantle conditions are similar. In the ridge regions, where the upper mantle is spreading, the molten material is of a lower density (thanks to its higher temperature), while in a subduction zone the density of the materials is higher because the materials being subducted are cooler. The difference in density between the spreading ridge and the subducting slab is (like that of an estuary) around 0.02 grams per cubic centimeter.

The plates are carried along on the surface by gravitational circulation flow from the hotter ridge to the cooler subduction zone, while the circulation loop is completed by a return flow down below.

The energy source for all this activity is heat. The Earth is a giant heat engine, and its core is the primary heat reservoir. Another product of the Earth's heat engine is its mantle plumes, the

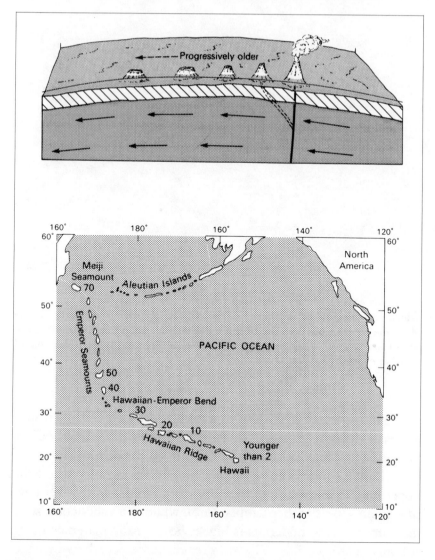

Above: How the Hawaiian volcanic island chain formed.
Below: The map shows ages, in millions of years, of volcanoes in the Hawaiian-Emperor chain. *Source:* Uyeda 1971.

conduits for the transfer of this heat from the core to the upper mantle. They are expressed on the Earth's surface in the form of hot spot volcanoes. These are a third type of volcano, the other two being those of the midocean ridges and those produced in regions of plate subduction.

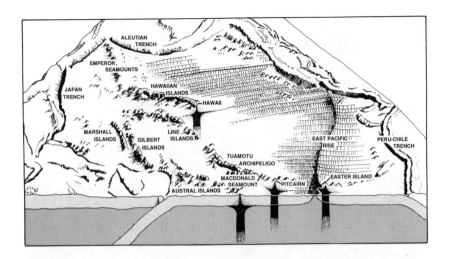

Bathymetric map of the Pacific Ocean, showing hot spot traces of the Hawaiian-Emperor, Tuamotu-Line, and Austral-Gilbert chains and their mantle plume sources. Also shown are the midocean spreading ridge of the East Pacific Rise and the subduction zones of the Japan, Aleutian, and Peru-Chile trenches. *Source:* **Adapted from Vink, Morgan, and Vogt 1985.**

The mantle plumes are, as far as anyone can tell, fixed geographically within the mantle itself, with the plates moving over the plume locations. As noted earlier, this is how the Hawaiian Islands came into being, extending west-northwest today from the island called Hawaii to the older, far more eroded island of Kauai. Beyond lie the Emperor Seamounts extending north-northwest, indicating that the Pacific Plate changed direction millions of years ago in its travels across the mantle plume. At present, some forty hot spots are active around the globe, including the Tuamotu Archipelago, Line Islands, Austral Islands, and Gilbert and Marshall Islands, all in the South Pacific (see above); the Walvis Ridge and Rio Grande Ridge in the South Atlantic; and the seamounts extending in a line from India's Deccan Traps to Reunion Island in the western Indian Ocean.

The mantle plumes, as we mentioned earlier, rise up from the boundary between the Earth's molten core and the solidified overlying mantle. This boundary is maintained because the density of the core material is much greater than that of the mantle. As heat

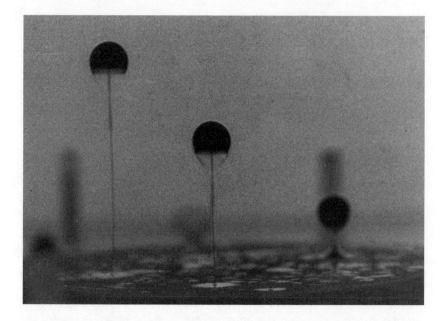

Multiple plumes arising from different points of the source region in a laboratory experiment. *Source:* **David Loper, Florida State University.**

escapes from the core by conduction, it warms the lowermost part of the mantle, which in turn becomes somewhat mushy. The heated, less dense material there then forms a plume, a conduit to carry the heated material up toward the surface. As the accumulation of hot liquid material moves up through the resistant mantle, it forms a mushroom-shaped head at the plume's top. The accompanying illustration shows this action from a laboratory experiment. When the head breaks through at the surface, a vast outpouring of volcanic material occurs over what in geologic terms is a relatively short period.

The whole process is cyclical. As a plume forms, the layer of heated material is exhausted upward, to be replaced by cooler material from the mantle. This is, in turn, heated enough to form another plume or replenish the existing one. Calculations of a theoretical nature suggest that the period required for the lowermost mantle material to heat up again enough to form one or more plumes and erupt at the Earth's surface is around 20 to 30 million years.

This theoretical cyclicity in mantle plume buildup can be considered in light of the known episodic nature of global volcanism. James Kennett of the University of California at Santa Barbara and others have found that over the past 30 million years there have been two periods of intense volcanic activity, with longer intervening intervals of relative quiescence. One of these two intense periods occurred from 14 to 16 million years ago; the other began some two million years ago and continues into the present.

Scientists have also noted a global synchronism of mantle plume activity. Peter Vogt of the U.S. Naval Research Laboratory found that hot spot discharges from Hawaii, Iceland, and other plume sites were high during the late Cretaceous to early Tertiary period, followed by a low and then a renewed increase up to the present. Superimposed on this broad trend are two additional highs, at 15 million years ago (the same as found by Kennett) and approximately 40 million years ago.

Vogt's findings were made in the early 1970s, and in 1972 he was one of the first to suggest a connection between the Deccan Trap eruptions in India and the mass extinctions that occurred near the K-T boundary. Then, as always, the Earth was dancing to its own tune, and life-forms either had to dance along with it or get off the stage.

We must now look in more detail at the kind of volcanism represented by the Deccan Traps. Nothing like it has occurred in the intervening 65 million years, but a similar eruption interval happened previously.

The Deccan volcanism presumably began with a violent sequence of explosive events and continued with a quieter outpouring of lava—in all, the estimated volume of lava was greater than one million cubic kilometers and covered a vast region of what is now India. But volcanism in K-T times was not confined to the Deccan Traps: it was global in nature. The Deep Sea Drilling Program cores record explosive K-T period volcanism in high-latitude Southern Hemisphere sites. In the southeastern Atlantic, drilling

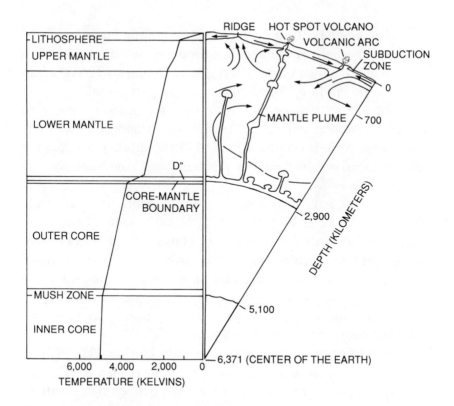

Mantle plume model. *Source:* **Courtillot 1990.**

sites on the Walvis Ridge show the ridge to be a hot spot mantle plume trace, seen now in volcanic sand layers lying across the K-T boundary.

A rather complete and clear story exists in the Walvis Ridge sites. From the K-T boundary layer backward in time into the late Cretaceous, there is a carbonate section interrupted by volcanic sand layers that extend more than five meters, or approximately 500,000 years, from magnetic polarity interval 30N into interval 29R. From the K-T boundary forward in time into the early Tertiary are fine volcanic ash sediments, ranging in a wonderful color array from red to green to brown and extending over another 500,000 years or so, from magnetic polarity interval 29R into 29N. (Nowhere else in the cores at these Walvis Ridge sites are such prominent color variations found.)

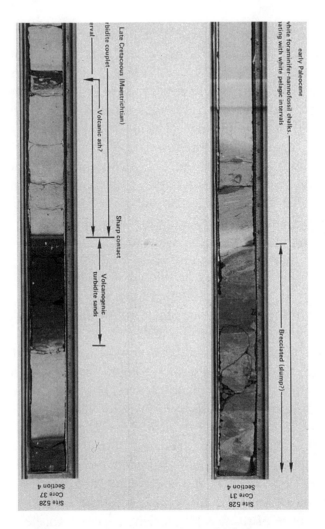

Photographs of cores from Deep Sea Drilling site 528 on the Walvis Ridge. Late Cretaceous core with volcanic black sands to the right. Early Tertiary core with fine volcanic ash layers to the left. *Source:* **Borella 1984.**

In addition, intense volcanic activity at and around K-T time occurred in North America. From the Carrizo, Ute, La Plata, and Elk mountains, to the southwest in the Four Corners (where Utah, Colorado, Arizona, and New Mexico meet), through the main Colorado mineral belt to the northeast, lies the Laramide magmatic trend. The Laramide trend is the most prominent magmatic development anywhere in North America. Potassium-argon radiometric dating, with an error range of one to three million years, gives it an age of 64 to 69 million years, confirming that it is a signature of the mountain-building episode that created the present Rocky Mountains in late Cretaceous and early Tertiary times.

At the same time that volcanic activity was creating the Laramide trend, a huge interior seaway that extended from the Gulf of New Mexico north to the Arctic Ocean during most of the Cretaceous period had begun to disappear. The disappearance began in the Albian age, and by the end of the Maastrichtian age, the seaway was totally gone. During the later Maastrichtian age, the level of the world's oceans began to drop as well, with North America's inland sea disappearing at an increasing rate.

Before taking yet a closer look at K-T times, an even longer look at the geologic record is now in order.

Mass extinctions, it has been noted elsewhere in this book, define the major divisions in the geologic timescale. A major extinction occurred at the Permian-Triassic boundary separating the Paleozoic era from the Mesozoic 250 million years ago, just as the K-T extinctions mark the end of the Mesozoic and the beginning of the Cenozoic 65 million years ago. Of the two, the P-Tr extinction was greater by far, and it is widely recognized as the most important biotic event in the entire Phanerozoic record. At that time, *70 to 80 percent* of all existing species disappeared, and *90 percent or more* of the world's marine species vanished. At K-T time, some 50 percent of existing species disappeared.

There is an exact, one-to-one correlation between environmental events associated with the P-Tr transition and those associated with the K-T transition. A major sea-level regression in the late Permian was followed by a transgression (or rise) in the early Triassic. They match almost exactly in magnitude, timing, and duration those that occurred at K-T time. A major episode of flood basalt volcanism occurred then also, again comparable in magnitude, timing, and duration to that of K-T time. The late Permian, early Triassic sea-level changes are apparent in exposed sites around the world and in drilling results in the Carnic

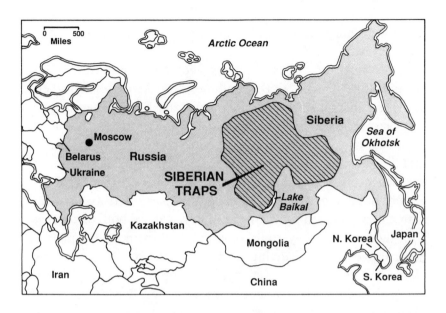

Siberian Traps location. *Source:* Browne 1992.

Alps in Austria: they confirm a fall of some 280 meters in sea level, followed by a rise in the Triassic.

The volcanism of the P-Tr transition period is best documented in the central Asian portion of Siberia, where the Siberian Traps were created by an outpouring of two to three million cubic kilometers of basalt. Uranium-lead dating of zircons there give an age of 244 to 252 million years, a nearly perfect match for the age of 247 to 255 million years assigned to the P-Tr transition by sedimentary studies. The eruptions in Siberia occurred over some 600,000 years, and except for the Deccan Traps, nothing like them has occurred since.

Interestingly (to look at a *minor* phenomenon in the midst of these titanic global events), drill cores from the Austrian sites showed two modest iridium anomalies—one of 1.65 ppb at the P-Tr boundary, and another of 2.3 ppb, some 39 meters above it. From the relative abundance of other elements, these anomalies are likely volcanic as opposed to meteoritic in origin. It is worth noting in this connection that the Siberian Traps contain one of the world's major reserves of the platinum group elements, including iridium.

Just how the cycles of sea-level regression and transgression around the P-Tr and K-T boundaries are related to the Siberian and Deccan volcanism remains a puzzle. Some scientists have suggested that sea levels fell as ocean waters were sucked up into polar ice, itself a result of atmospheric cooling from the aerosols that the volcanism injected into the stratosphere. But the timing of the two events does not appear to accord with this suggestion. Others have suggested that changes in plate density associated with a new plate-rifting event could have caused both phenomena. Our guess is that they could have resulted from *vertical* plate tectonic movements occurring as adjustments to new or renewed plume sources rising through the mantle.

Whatever the explanation, these two events—intense volcanism and a sea-level regression-transgression cycle—*did occur.* They are facts, major facts, along with the major facts of mass extinctions that coincided with them.

Now let us take a closer look at the actual events—environmental changes and extinctions—that occurred in K-T times.

Three major environmental changes occurred in K-T time, and in this sequence: the loss of North America's inland seaway, the peak in sea-level regression (which is to say, the nadir of the sea level), and the Deccan volcanism. Three major extinctions occurred in this long period as well, and in this sequence: first, the dinosaurs; second the shallow-water shellfish; and third, the ocean plankton.

When the inland seaway of North America began to disappear, the land on its shores also began to change. The near shores had been lush tropical places with extensive coal swamps, while farther inland was a temperate floral environment. All this changed as the inland sea shrank. The lush vegetation diminished, following the shrinking shoreline, and eventually the entire region became an uplifted arid environment, not unlike much of the American Southwest today. The lush swampy areas and the rest of the densely vegetated region had been dinosaur habitat, and as it shrank, so did the dinosaur population, presumably until it collapsed altogether. Western North America, as noted, was the last refuge for dinosaurs, their disappearance elsewhere in the world having already occurred.

Beginning some 69 million years ago, sea levels dropped, reaching a low point at the end of the Cretaceous, in an overall regression of as much as 100 meters. Today, shallow waters are an extensive feature of the world's coastlines, but at depths of 100 meters, there is a sharp break in the bottom topography. This is the edge of the continental shelves, where the seafloor plunges rapidly toward the deep ocean basins. If the Cretaceous coasts and coastal waters were anything like what exists today, a drop of 100 meters in sea level would have removed virtually all the shallow-water regions. For an extended period of time, virtually no habitat would have existed in which shallow-water shellfish could survive. One would expect such populations to collapse, and a significant turnover between species before and afterward. This is, exactly what happened among shellfish.

From the geologic record of the Deccan volcanism and from measurements of recent volcanic emissions, we have a clear picture of other environmental changes that occurred at K-T time. At its peak, K-T volcanism may have been a hundred times greater than what we experience today—that is, it would be something like a Mount St. Helen's (or bigger) happening every day of every year.

During peak periods, chlorine emissions would have been 110 times those that are today estimated to cause an 8 percent reduction in the protective ozone layer. There would have been increased ultraviolet radiation. Sulfur dioxide emission would have been fourteen times what arises from present-day fossil fuel emissions. That is a global average: acid rain for any given location or time could have substantially exceeded this value. Sulfur dioxide returning to the oceans as acid rain would have been sufficient to reduce the pH of surface waters from their normal 8.2 to 7.6 and maybe lower. Lowered pH means greater acidity, which reduces the availability of the carbonate ion needed in building calcareous plankton shells. Such a pH change would be long last-ing—a particle of water has a lifetime of some fifty years in the 100-meter realm of surface waters.

Volcanic aerosols, particularly sulfur dioxide, have a strato-spheric residence of a year or so, and their effect from large vol-canic eruptions is to cool the Earth's surface temperature. On the other hand, the associated carbon dioxide emissions would lead to a greenhouse effect and global warming. The combined effects of the two—sulfur dioxide and carbon dioxide—remain moot since they tend to cancel each other out.

The crux for any late Cretaceous and K-T scenario of envi-ronmental events is how well it explains those floral and faunal changes that are clear in the geologic record. Not only must it pro-vide a model that explains species extinctions, it must also account for species that survived the K-T transition with little or no change. After all, 50 percent of life-forms did survive.

If the last dinosaurs disappeared in North America along with their habitat, and if most shallow-water shellfish disappeared along with theirs, what happened to the plankton—both the ani-mal and plant forms?

Today, foraminifera (faunal plankton) are inhibited below a pH of 7.6 to 7.8; modern phytoplankton do not survive below a pH of 7.0 to 7.3. If planktonic forms of K-T times were similarly affected, the changes in pH from the volcanism of the period would have led to extinctions of both types of plankton, with the foraminiferal extinctions preceding the phytoplankton extinctions. Further, among modern plankton, smaller forms survive better in a lower pH environment than larger ones. Indeed, given any change away from a stable environment, smaller forms do better than larger ones. The geologic record shows this to have been the case at the K-T boundary, when the larger, more ornate plank-tonic forms of the Cretaceous were replaced by the Tertiary's smaller, simpler forms. The major effects of pH change would have been seen in tropical and temperate surface waters, and any global cooling would have favored hardier forms from higher lati-tudes. The earliest Tertiary replacement forms—both plant and animal—are of a higher-latitude, colder-water type.

Least affected among all the planktonic forms across the K-T boundary were the *dinoflagellates,* such as the phytoplankton that cause red tides. Dinoflagellates don't depend on carbonate. Today, in freshwater lakes with pH levels of 5 to 4, dinoflagellates are among the few forms that can survive, so it is not surprising to find dinoflagellates doing better at K-T times than the carbonate-dependent forms of plankton.

By the same token, benthic (or bottom-dwelling) foraminifera that inhabited the deep ocean depths show little change at the K-T boundary. Changes in pH would have been greatest in surface waters, especially if Cretaceous seas were more stratified than today's oceans, as would seem to have been the case. Environmental changes at the time would have been smaller, and had fewer effects, on the benthic foraminifera.

Practically no changes are to be seen across the K-T boundary among placental mammals, birds, amphibians, and reptiles. (As noted, dinosaurs and four other reptilian orders had already died out before the end of the Cretaceous.) Presumably, among such a diverse group of surviving organisms, the protective elements were equally diverse. Different factors may be speculated about for different species—for example, large population sizes, wide temperature tolerances, and nocturnal habits, and subaquatic and subterranean habitats that gave protection from the increased ultraviolet radiation.

Freshwater fish at K-T time showed selective extinctions, possibly related to acid rain. In a modern Canadian lake that has undergone severe changes in pH from acid rain, walleye and trout disappear at a pH of 5.8 to 5.2; northern pike and white sucker at 5.2 to 4.7, while yellow perch and lake herring survive a pH of less than 4.7. There is no reason not to expect a similar pattern of extinction at K-T time. Similar floral effects from acid rain also occur today—with a gradient of sensitivity in the northeastern United States and adjacent Canada ranging from highly sensitive white pine and white birch, to red pine and balsam fir, to more but not totally resistant red cedar and red maple. Why wouldn't late Cretaceous species show some similar range of sensitivities?

Such are the events—titanic global events, lesser regional events, and local events down even to the microscopic level—that are known to have occurred when the Cretaceous period ended and the Tertiary began. They are not quite so dramatic as a rain of meteorites, but they *are* known to have occurred. It would probably have been pretty scary living then, particularly if someone could have actually seen what was taking place over the long periods before, during, and ahead of his or her own tiny instant of existence in geologic time.

Today the geologic record provides us with a time machine to visit those earlier times and their great upheavals. Today we need no time machine to see what our own effects on life-forms, including our own, most likely will be. As for the Earth's intentions, those we must simply await, knowing we have as little likelihood of influencing them as the dinosaurs and shellfish and plankton had at the end of the Cretaceous.

Those particular events have all pretty much died down by now, but at least one memento remains from those times that still causes a bit of havoc today. We mention it not because of its Earth-shattering importance but merely as a reminder of the continuity of things on a planet like the Earth. It is *selenium*.

Here we return to the volcanism associated with the Laramide mountain-building episode in western North America that provided us with the Rocky Mountains. We know less about the specifics of this volcanism than we do about the Deccan Traps, but we do know that it lasted over a longer period of time, with eruptions at various locales that were, all together, probably of smaller total magnitude. But rains brought volcanic dust and other airborne particles from this regional disturbance into the shallow seas of the Western Interior Seaway, where they collected on the bottom in succeeding sedimentary layers. Some of these sedimentary strata—Upper Cretaceous shales—appear in outcrops around the Great Plains and beneath the soil as well. The selenium concentrations in some of these shales have extremely high values—in the range of 10 to 20 parts per million (ppm).

Now, when William Zoller analyzed airborne particles from Kilauea, the element that showed the greatest enhancement was selenium—an enhancement, in fact, of *40 million.* Also, in the dust bands in the Antarctic from relatively recent eruptions, the selenium concentrations range from 10 to 20 ppm. The K-T sections have similar enhancements, from 10 to 20 ppm.

Herbs and grasses grow with great facility in soils on the rock outcrops of the Great Plains. Many of these plants—particularly one called *Astragalus* by scientists and known otherwise as locoweed—accumulate selenium from the soils into their own tissues, where it reaches levels of 500 to 1,000 ppm—levels that are highly toxic.

The first documented account of the effects of this Cretaceous-age selenium was given by T. C. Madison, a surgeon attached to a U.S. Cavalry unit stationed at Fort Randall, North Dakota. Dr. Madison had this to say:

> Four companies of the second dragoons arrived at this post about the 10th of August, 1856, one squadron from Fort Lookout and one from Big Sioux River. . . . The four companies encamped on the lower side of the dry ravine, separating the dragoon and infantry camps. About the 20th of August the disease commenced simultaneously in all four companies, and many horses died, not however, until the lapse of weeks and months. The following symptoms were observed: first, that, among the remount horses from below, there was a sort of catarrh, or distemper, with running at the nose, and among all horses a swelling of the skin of the throat and jaw; also inflammation, swelling and suppuration at the point where the hoof joins the skin, the hoof, in a measure detaching itself, and a new one forming in its place. These were also accompanied by loss of manes and tails. The appetite was uniformly good, but, from extreme tenderness of the feet, they were unable to move about in search of food, and it appears that at the time they were entirely dependent upon grazing, there being no forage at the post for issue. . . . A few mules and Indian ponies were similarly affected. The

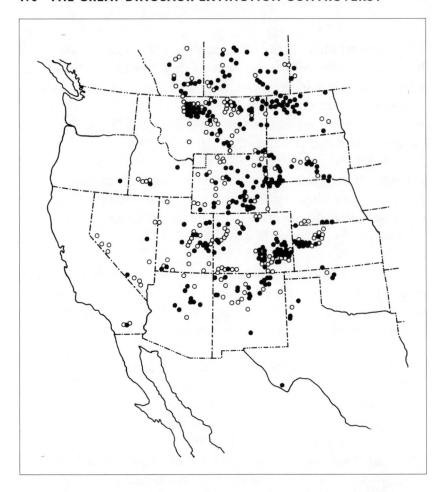

Distribution of seleniferous vegetation in the western United States and Canada. Each open dot represents the place of collection of a plant specimen containing 50 to 500 ppm selenium; each solid dot, specimens containing more than 500 ppm. *Source:* **Rosenfeld and Beath 1964.**

acclimated suffered equally with the unacclimated. No treatment was effectual, or afforded permanent relief.

Neither Dr. Madison nor anyone else at the time had any idea that what was happening to their livestock had anything to do with selenium; nor could anyone at the time have made any connection between their livestock problems and the demise of the dinosaurs, a few remains of which had come to light only a few decades earlier.

In any event, other such poisonings followed as the Great Plains began to be used for cattle grazing. Some thought the cause was saline waters and salt crusts—hence the name "alkali disease" became attached to one form of the poisoning. It was not until 1933 that selenium was identified as the culprit.

Seleniferous plants produce either acute or chronic poisoning in livestock. Acute poisoning usually arises from a single feeding on plants with high selenium concentrations—500 ppm or more. In many cases death follows within a few hours. Plants with lower concentrations can cause two chronic forms of poisoning. In the *blind staggers,* the animal wanders aimlessly in circles and has little desire to eat or drink, followed by collapse and then paralysis prior to dying. In *alkali disease,* the animal shows a lowering of vitality, lameness, loss of hair, and hoof lesions and deformity. In cases of chronic selenium poisoning, if an animal is simply taken off seleniferous feed, it will usually recover. Stockmen have learned to fence off areas of selenium-rich soils from their livestock, and to remove locoweed either by uprooting it or spraying it with herbicides.

Thus are the last active results of the titanic events of the K-T boundary dealt with—not with a bang but an agricultural program.

AFTERWORD
Pathological Science

THE ALVAREZ HYPOTHESIS HAS COLLAPSED UNDER THE weight of accumulated geologic and other evidence to the contrary, as well as from an increasingly obvious absence of scientific evidence proffered in its support. An important question remains, however: Is this story simply an unfortunate but unique interval in the history of science (an anomaly, one is tempted to say), or does it bear useful lessons for laymen, science journalists, and even scientists themselves? In fact, it carries several such lessons.

Philosophers of science have tried to distinguish the criteria that characterize a new hypothesis that eventually becomes accepted as a working model (or paradigm), and one that is found lacking and is in due course discarded. In his book *The Methodology of Scientific Research Programmes,* Imre Lakatos, an eminent philosopher of science, gives a clear discussion of the criteria of what he calls a progressive research program and a degenerative one. We quote extensively here from his introduction; the italics are added for emphasis.

Now, Newton's theory of gravitation, Einstein's relativity theory, quantum mechanics, Marxism, Freudianism, are all research programmes, each with a characteristic hard core stubbornly defended, each with its more flexible protective belt and each with its elaborate problem-solving machinery. *Each of them, at any stage of development, has unsolved problems and undigested anomalies.* All theories, in this sense, are born refuted and die refuted. But are they equally good? Until now, I have been describing what research programmes are like. *But how can one distinguish a scientific or progressive programme from a pseudoscientific or degenerating one?*

Contrary to Popper [Karl Popper, another philosopher of science], the difference cannot be that some are still unrefuted, while others are already refuted. When Newton published his *Principia*, it was common knowledge that it could not properly explain even the motion of the moon; in fact, lunar motion refuted Newton. Kaufmann, a distinguished physicist, refuted Einstein's relativity theory in the very year it was published. But all the research programmes that I admire have one characteristic in common. *They all predict novel facts, facts which had been either undreamt of, or have indeed been contradicted by previous or rival programmes.* In 1686, when Newton published his theory of gravitation, there were, for instance, two current theories concerning comets. The more popular one regarded comets as a signal from an angry God warning that He will strike and bring disaster. A little known theory of Kepler's held that comets were celestial bodies moving along straight lines. Now according to Newtonian theory, some of them moved in hyberbolas or parabolas never to return; others moved in ordinary ellipses. Halley, working in Newton's programme, calculated on the basis of observing a brief stretch of a comet's path that it would return in seventy-two years' time; he calculated to the minute when it would be seen again at a well-defined point of the sky. This was incredible. But seventy-two years later, when both Newton and Halley were long dead, Halley's comet returned exactly as Halley predicted. Similarly, Newtonian scientists predicted the existence and exact motion of small planets which had never been observed before. Or let us take Einstein's programme. This programme made the stunning prediction that if one measures the distance between two stars in the night and if one measures the distance between them during the day (when they are visible during an eclipse of the sun), the two measurements will be different. Nobody had thought to make such an observation before Einstein's programme. *Thus, in a progressive research programme, theory leads to the discovery of hitherto unknown novel facts. In degenerating programmes, however, theories are fabricated only in order to accommodate known facts.* Has, for instance, Marxism ever predicted a stunning novel fact successfully? Never! It has some famous unsuccessful predictions. It predicted the absolute impoverishment of the working class. It predicted that the first socialist revolution would take place in the industrially most developed society. It predicted that socialist societies would be free of revolutions. It predicted that there will be no conflict of interests between socialist countries. Thus the early predictions of Marxism were bold and stunning but they failed. Marxists explained all their failures: they explained the rising living standards of the working class by devising a theory of imperialism; they even explained why the first socialist revolution occurred in industrially backward Russia. They "explained" Berlin 1953, Budapest, 1956, Prague, 1958. They "explained" the Russian-Chinese conflict. But their auxiliary hypotheses were all cooked up after the event to protect Marxian theory from the facts. The Newtonian programme led to novel facts; the Marxian lagged behind the facts and has been running fast to catch up with them.

To sum up. The hallmark of empirical progress is not trivial verifications: Popper is right that there are millions of them. It is no success for Newtonian theory that stones, when dropped, fall towards the earth, no matter how often this is repeated. But so-called "refutations" are not the hallmark of empirical failure, as Popper has preached, since all programmes grow in a permanent ocean of anomalies. *What really counts are dramatic, unexpected, stunning predictions: a few of them are enough to tilt the balance; where theory lags behind the facts, we are dealing with miserable degenerating research programmes.*

Now, how do scientific revolutions come about? If we have two rival research programmes, and one is progressing while the other is degenerating, scientists tend to join the progressive programme. This is the rationale of scientific revolutions. But while it is a matter of intellectual honesty to keep the record public, it is not dishonest to stick to a degenerating programme and try to turn it into a progressive one.

As opposed to Popper the methodology of scientific research programmes does not offer instant rationality. One must treat budding programmes leniently: programmes may take decades before they get off the ground and become empirically progressive. Criticism is not a quick kill, by refutation. Important criticism is always constructive: there is no refutation without a better theory. Kuhn [Thomas Kuhn, another scholar of science] is wrong in thinking that scientific revolutions are sudden, irrational changes in vision. The history of science refutes both Popper and Kuhn: on close inspection both Popperian crucial experiments and Kuhnian revolutions turn out to be myths: *what normally happens is that progressive research programmes replace degenerating ones.*

Further into the book Lakatos has the following brief paragraph:

The time-honoured empirical criterion for a satisfactory theory was agreement with observed facts. *Our empirical criterion for a series of theories is that it should produce new facts. The idea of growth and the concept of empirical character are soldered into one.*

And with regard to the replacement of the Bohr theory of the atom by wave mechanics, he says the following:

But temerity in proposing wild inconsistencies did not reap any more rewards. *The programme lagged behind the discovery of "facts." Undigested anomalies swamped the field. With ever more sterile inconsistencies and ever more ad hoc hypotheses, the degenerating phase of the research programme had set in: it started—to use one of Popper's favorite phrases—"to lose its empirical character."* Also many problems, like the theory of perturbations, could not even be expected to be solved within it. A rival research programme soon appeared: wave mechanics. Not only did the new programme, even in its first version (de Broglie, 1924), explain Planck's and Bohr's quantum conditions; it also led to an exciting new fact, to the Davisson-Germer experiment. In its later, ever more sophisticated versions it offered solutions to problems which had been completely out of the reach of Bohr's research programme, and explained the *ad hoc* later theories of Bohr's programme by theories satisfying high methodological standards. Wave mechanics soon caught up with, vanquished and replaced Bohr's programme.

De Broglie's paper came at a time when Bohr's programme was degenerating. But this was mere coincidence. One wonders what would have happened if de Broglie had written and published his paper in 1914 instead of 1924.

This set of criteria, developed from the philosophy of science, can be used to assess whether a given research program may be considered progressive or degenerative. It can help us determine which of two prevailing hypotheses—impact or volcanic—may be on the right track toward explaining, at least in part, the extinctions that occurred at and around K-T time, and which may be on the wrong track. A progressive research program will lead to the prediction of new and unusual facts that had not been considered previously or that had even been contradicted by previous hypotheses. Its stunning and unexpected predictions lead to the discovery of hitherto unknown and novel facts. In degenerative research programs, however, theories are fabricated only to accommodate known facts. New corollaries and hypotheses are added on an ad hoc basis as facts that are not answered by the original hypothesis emerge.

Prior to the Alvarez impact hypothesis in 1980, it was presumed that the only significant source for iridium on the Earth's surface had to be meteoritic materials. This was the cornerstone for the original hypothesis, and it was a logical deduction at the time. But if the volcanic explanation for extinctions is correct, the unavoidable conclusion is that, contrary to common wisdom, iridium has to be associated with volcanic eruptions. Albeit serendipitously, William Zoller found that it was, and later investigators confirmed it. Enhanced iridium levels occur in the fine particles emitted from volcanoes, and at K-T sites it is in fine clay particles that one finds iridium enhanced.

In support of the Alvarez hypothesis, microscopic deformation features in quartz grains at K-T sections, found by Bruce Bohor in 1984, were taken as further evidence of an impact. The supposition, shared by many, was that such features could be explained only by impact processes. But again, if the volcanic hypothesis is correct, the unavoidable conclusion is that, again contrary to accepted wisdom, such deformation features have to be found in association with volcanic eruptions. That they are was confirmed by Neville Carter's studies in 1986.

Both discoveries—iridium and PDFs—fall under the category of novel and undreamed-of findings that contradict common wisdom, and both were the result of the volcanic hypothesis—in this sense, a progressive research program. Like the progressive research programs Lakatos describes, the volcanic hypothesis has its own "unsolved problems and undigested anomalies." A major one awaiting possible solution is the relationship, if one exists, between the flood basalt volcanism and the sea-level regression-transgression cycle that was apparent at both the K-T and P-Tr transitions.

What sort of research program do the impactors have? Soon after 1980, it became evident that an extended and sequential record of extinctions cannot be easily explained by a single event of either terrestrial or extraterrestrial origin. How did the impactors adjust to this development? They called on six to ten cometary impacts over three million years, as opposed to a single asteroid impact at the K-T boundary. This new and stunning suggestion did not result in any way as a prediction from the original hypothesis. It was merely an ad hoc attempt to adjust to the observed extinction record once its nature had been forcefully enough pointed out to the impactors.

Another problem for the impactors came in the form of geophysical and geochemical data concerning oceanic and terrestrial impact sites. The PDFs suggest a continental target, but arsenic and antimony suggest an oceanic one. How did the impactors resolve this discrepancy? They invoked yet another scenario, involving not one big K-T meteorite but several smaller ones falling over a period of a year or so, some hitting continents, others hitting oceans.

Both of these adjustments—multiple impacts over a year and multiple impacts over three million years—were fabricated to accommodate known facts. Neither sprang in any way from the original hypothesis; neither led to any novel facts. That is to say, they are characteristic of a degenerative research program. And as discussed in Chapter 9, neither has stood up to the test of scientific scrutiny.

Scientists, contrary to the popular archetype, are often quite unobjective in the pursuit of truth. The seventeenth-century philosopher John Locke wrote, "It is in man's power to content himself with the proofs he has, if they favor the opinion that suits with his inclinations or interest, and so stop from further research." Another set of criteria—in a sense, "symptoms"—can help in "diagnosing" an emerging scientific research program or hypothesis as degenerative. These symptoms point to what was called *pathological science* by the late Irving Langmuir, a Nobel Prize–winning chemist from General Electric, in a lecture that he gave in 1953; they were expanded upon in 1992 by Denis Rousseau of AT&T Bell Laboratories. Though developed in the context of physics and chemistry, the applicability of Langmuir's symptoms to the Alvarez hypothesis is nearly uncanny.

The first symptom of pathological science is that *the maximum effect is produced by a causative agent of barely detectable intensity.* The present debate started with the Alvarez findings on iridium, an extremely rare element in the Earth and about as common as those rare visitors to the Earth, meteorites. Prior to 1980, the literature on the geochemistry of iridium was itself sparse, much of it only an adjunct to mineral exploration and the mining of the platinum group of elements.

Second, *the effect is of a magnitude that remains close to limit of detectability*. Even a substantial iridium anomaly at the K-T boundary is measured in the range of five to ten parts per *billion*. Iridium is at best a minor part of the overall environmental conditions of K-T times—especially when compared with sea-level drops of 100 meters and the spewing out of millions of cubic kilometers of magma from volcanoes, along with vast quantities of volatiles into the skies. In comparable terms, making note of iridium resembles making note of two particularly odd individuals in the total population of the United States.

Third, *the investigators claim great accuracy of measurement*. With the potential importance of iridium in unraveling the causes of extinctions, detection techniques have been developed to determine levels of parts per *trillion*.

Fourth, *fantastic theories that contradict experience are presented*. Six thousand years of recorded human civilization have witnessed a variety of natural catastrophes. Accurate figures do not exist, but a reasonable estimate of fatalities from earthquakes alone is in the range of tens to hundreds of millions. Fatalities from volcanic eruptions, such as that at Santorini in 1470 B.C. in the eastern Mediterranean (which evidently wiped out the Minoan civilization on Crete) and at Krakatoa in 1883, along with flooding and other natural disasters, probably have the same fatality range as earthquakes. Yet in all this time, not one fatality has ever been reported from a meteorite impact.

Nevertheless, a warning was implicit in the Alvarez hypothesis: It could happen again. Astronomers successfully went about securing large amounts of funding for sentinel duty. To the extent that this furor distracted humanity's attention from the ways in which it is itself inaugurating a period of mass extinctions, the Alvarez hypothesis has been not merely pathological science but dangerous to boot.

Fifth, *criticisms are met with ad hoc explanations presented on the spur of the moment.* This goes back to the Lakatos criterion for recognizing a degenerative research program. Six to ten cometary impacts, rather than a single asteroid impact, over a period of three million years, were proposed in order to explain the observed sequential and extended extinction record. Several impacts over a year or so, on both oceanic and continental terrains, were proposed in order to explain the presence of both anomalous amounts of arsenic and antimony and quartz grains with planar deformation features.

Sixth, *the ratio of supporters to critics rises to somewhere near 50 percent and then falls gradually to oblivion.* This is about the way things have gone. A number of scientists within the planetary geology community were initially attracted to the asteroid impact hypothesis, which called attention to the potential importance of extraterrestrial events in shaping Earth history. By and large, however, paleontologists, sedimentologists, and stratigraphers did not think much of the hypothesis, for it could not account for the sequential and extended nature of the extinctions in the geological record. As we have seen, an international survey taken in 1984 showed that those who considered an asteroid impact to be the causative agent for the extinctions were as follows: 9 percent of British paleontologists, 10 percent of Soviet geoscientists, 14 percent of German paleontologists, 16 percent of Polish geoscientists, and 31 percent of American geophysicists. We know of no more recent surveys, but the percentage of impact supporters has not, as yet, fallen to zero.

In all, *pathological science arises from self-delusion.* Sometimes a scientist who believes he or she is acting methodically has really lost all objectivity. This symptom can be broken down into several components, all of which were transparently present in the various searches for a meteorite crater in the Gulf of Mexico and Caribbean region, undertaken to vindicate the impact theory to at least some degree.

The first component is *hurried, careless, or incomplete work.* The excitement over the Big Boulder Bed in Cuba and the insufficient care with which drilling sites 536 and 540 were studied by revisionists are examples of this carelessness.

The second component is *the invention of a new science to fit the observations*. In this case, a new geology was invented at the Haiti geologic site and at the Mimbral outcrop in Mexico.

The third component is *ignoring facts that do not agree with preconceived notions*. This component applies to all four proposed impact sites, including Chicxulub.

Now that we have been forewarned, perhaps we can all be a bit more attentive to the evidence. There is no guarantee that the human tendency to be satisfied with proofs that support one's own inclinations will cease to apply in the affairs of science. Degenerative research programs will get started as surely as water seeks its own level. Happily, the very processes of sciences have so far tended to overcome these human shortcomings over time. We do make progress.

BIBLIOGRAPHY

ONE *The Day of the Meteorite*

ASTEROID IMPACT HYPOTHESIS AND ITS RECEPTION

Alvarez, L. W., W. Alvarez, F. Asaro, and H. V. Michel. 1980. Extraterrestrial cause for the Cretaceous-Tertiary extinction. *Science* 208:1095–108.

Clemens, E. S. 1986. Of asteroids and dinosaurs: The role of the press in the shaping of scientific debate. *Social Studies of Science* 16:421–56.

DINOSAUR EXTINCTION THEORY PRIOR TO 1980

Benton, M. J. 1989. Scientific methodologies in collision: The history of the extinction of the dinosaurs. *Evolutionary Biology* 24:371–400.

CROSS-FERTILIZATION IN SCIENCE

Jastrow, R. 1983. The dinosaur massacre: A double-barreled mystery. *Science Digest* (September): 51–53, 109.

Krishtalka, L. 1989. *Dinosaur Plots and Other Intrigues in Natural History.* New York: Avon Books.

Officer, C., and J. Page. 1993. *Tales of the Earth.* New York: Oxford University Press.

Wilford, J. N. 1986. *The Riddle of the Dinosaur.* New York: Alfred A. Knopf.

TWO *Meteorites and Comets*

COMETS AND ASTEROIDS

Browne, M. W. 1994. Comet that shook Jupiter may reveal inner secrets. *New York Times*, 26 July:C1, C11.

Chapman, R. D. and J. C. Brandt. 1984. *The Comet Book*. Boston: Jones and Bartlett.

Etter, R., and S. Schneider. 1985. *Halley's Comet, Memories of 1910*. New York: Abbeville Press.

Guillemin, A. 1875. *Les Comètes*. Paris: Libraire Hachette.

Heide, F. 1964. *Meteorites*. Chicago: University of Chicago Press.

Lancaster-Brown, P. 1985. *Halley and His Comet*. London: Blandford Press.

ASTEROID IMPACT RECORD ON EARTH

Benton, M. J. 1993. Late Triassic extinctions and the origin of the dinosaurs. *Science* 260:769–70.

Grieve, R.A.F. 1980. Impact cratering on Earth. *Scientific American* 262, no. 4:66–73.

Grieve, R.A.F. 1987. Terrestrial impact structures. *Episodes* 10:86.

Hoyt, W. G. 1987. *Coon Mountain Controversies*. Tucson: University of Arizona Press.

Officer, C., and J. Page. 1993. *Tales of the Earth*. New York: Oxford University Press.

EFFECTS OF ASTEROID IMPACTS

Aubry, M.-P., F. M. Gradstein, and L. F. Jansa. 1990. The late early Eocene Montagnais bolide: No impact on biotic diversity. *Micropaleontology* 36:164–72.

Hessig, K. 1986. No effect of the Ries impact event on the local mammal fauna. *Modern Geology* 10:171–79.

Lewis, H. R. 1990. *Technological Risk*. New York: W.W. Norton.

Officer, C. B., and R.A.F. Grieve. 1986. The impact of impacts and the nature of nature. *Eos* 33:633–37.

Poag, C. W., L. J. Poppe, and J. A. Commeau. 1993. The Tom's Canyon "impact crater," New Jersey OCS: The seismic evidence. *Geological Society of America Abstracts* 25:A378.

Poag, C. W., D. S. Powars, L. J. Poppe, and R. B. Mixon. 1994. Meteoroid mayhem in Ole Virginny: Source of the North American tektite strewn field. *Geology* 22:691–94.

Sawatsky, H. B. 1975. Astroblemes in Williston Basin. *American Association of Petroleum Geologists Bulletin* 59:694–710.

Weissman, P. R. 1985. Terrestrial impactors at geological boundary events: Comets or asteroids. *Nature* 314:517–18.

THREE *A Brief History of Dinosaurdom*

MESOZOIC ENVIRONMENT

Savin, S. M. 1982. Stable isotopes in climatic reconstructions. In W. H. Berger and J. H. Crowell, eds., *Climate in Earth History*. Washington: National Academy of Sciences.

Uyeda, S. 1971. *The New View of the Earth*. San Francisco: W.H. Freeman.

MESOZOIC LIFE

Buffetaut, E. 1990. Vertebrate extinctions and survival across the Cretaceous-Tertiary boundary. *Tectonophysics* 171:337–45.

Carroll, R. L. 1988. *Vertebrate Paleontology and Evolution*. San Francisco: W.H. Freeman.

Colbert, E. H. 1969. *Evolution of the Vertebrates*. New York: John Wiley.

Levin, H. L. 1991. *The Earth through Time*. 4th ed. Orlando: Saunders College Publishing.

Norman, D. 1985. *The Illustrated Encyclopedia of Dinosaurs*. New York: Crescent Books.

Schopf, T.J.M. 1982. Extinction of the dinosaurs: A 1982 understanding. *Geological Society of America Special Paper* 190:415–22.

Sloan, R. E., J. K. Rigby, Jr., L. M. Van Valen, and D. Gabriel. 1986. Gradual dinosaur extinction and simultaneous ungulate radiation in the Hell Creek formation. *Science* 232:629–33.

Sullivan, R. M. 1987. A reassessment of reptilian diversity across the Cretaceous-Tertiary boundary. *Natural History Museum of Los Angeles County Contributions in Science* 391:1–26.

Unwin, D. M. 1987. Pterosaur extinction: Nature and causes. *Société Géologique de France Mémoires* 150:105–11.

Van Valen, L., and R. E. Sloan. 1977. Ecology and extinction of the dinosaurs. *Evolutionary Theory* 2:37–64.

FOUR *Paleonecrology*

AGE OF THE EARTH

Burchfield, J. D. 1975. *Lord Kelvin and the Age of the Earth*. New York: Science History Publications.

Hallam, A. 1989. *Great Geological Controversies*. 2d ed. Oxford: Oxford University Press.

Levin, H. L. 1991. *The Earth through Time*. 4th ed. Orlando: Saunders College Publishing.

RADIOMETRIC AND PALEOMAGNETIC AGE DETERMINATIONS

Faure, G. 1977. *Principles of Isotope Geology*. New York: John Wiley.

Garland, G. D. 1979. *Introduction to Geophysics*. 2d ed. Philadelphia: W.B. Saunders.

Phillips, J. 1860. *Life on the Earth: Its Origin and Succession*. Cambridge, Mass.: Macmillan.

Stacy, F. D. 1977. *Physics of the Earth*. 2d ed. New York: John Wiley.

PALEONECROLOGY

Arthur, M. A., and A. G. Fischer. 1977. Upper Cretaceous-Paleocene magnetic stratigraphy at Gubbio, Italy. I. Lithostratigraphy and sedimentology. *Geological Society of America Bulletin* 88:367–71.

Kaufman, E. G. 1984. The fabric of Cretaceous marine extinctions. In W. A. Berggren and J. A. Van Couvering, eds. *Catastrophes and Earth History*. Princeton: Princeton University Press.

Signor, P. W., and J. H. Lipps. 1982. Sampling bias, gradual extinction patterns and catastrophes in the fossil record. *Geological Society of America Special Paper* 190:291–96.

Ward, P. D., W. J. Kennedy, K. G. MacLeod, and J. F. Mount. 1991. Ammonite and inoceramid bivalve extinction pattern in Cretaceous/Tertiary boundary sections of the Biscay region (southwestern France, northern Spain). *Geology* 19:1181–84.

Wiedmann, J. 1969. The heteromorphs and ammonoid extinction, *Biology Revue* 44:563–602.

FIVE *Early Dissent*

KRAKATOA COMPARISON

Alvarez, L. W., W. Alvarez, F. Asaro, and H. V. Michel. 1981. Asteroid extinction hypothesis. *Science* 211:654–56.

Kent, D.V. 1981. Asteroid extinction hypothesis. *Science* 211:649–50.

Ninkovich, D., R.S.J. Sparks, and M. T. Ledbetter. 1978. The exceptional magnitude and intensity of the Toba eruption, Sumatra: An example of the use of deep-sea tephra layers as a geological tool. *Bulletin Volcanologique* 41:286–98.

Simkin, T., and R. S. Fiske. 1983. *Krakatoa 1883*. Washington: Smithsonian Institution Press.

Stommel, H., and E. Stommel. 1983. *Volcano Weather*. Newport: Seven Seas Press.

DINOSAUR AND OTHER EXTINCTIONS AT THE K-T TRANSITION

Archibald, J. D., and W. A. Clemens. 1982. Late Cretaceous extinctions. *American Scientist* 70:377–85.

Keller, G. 1989. Extended period of extinctions across the Cretaceous/ Tertiary boundary in planktonic foraminifera of continental shelf sections: Implications for impact and volcanism theories. *Geological Society of America Bulletin* 101:1408–19.

Lerbeckmo, J. F., M. E. Evans, and H. Baadsgaard. 1979. Magneto-stratigraphy, biostratigraphy and geochronology of Cretaceous-Tertiary boundary sediments, Red Deer Valley. *Nature* 279: 26–30.

Schopf, T.J.M. 1982. Extinction of the dinosaurs: A 1982 understanding. *Geological Society of America Special Paper* 190:415–22.

Sloan, R. E., J. K. Rigby, Jr., L. M. Van Valen, and D. Gabriel. 1986. Gradual dinosaur extinction and simultaneous ungulate radiation in the Hell Creek formation. *Science* 232:629–33.

Thierstein, H. R., and H. Okada. 1979. The Cretaceous/Tertiary boundary event in the North Atlantic. *Initial Reports of the Deep Sea Drilling Project* 43:601–16.

Williams, M. E. 1994. Catastrophic versus noncatastrophic extinction of the dinosaurs: Testing, falsibility, and the burden of proof. *Journal of Paleontology* 68:183–90.

Zinsmeister, W. J., R. M. Feldmann, M. O. Woodburne, and D. H. Elliott. 1989. Latest Cretaceous/earliest Tertiary transition on Seymour Island, Antarctica. *Journal of Paleontology* 63:731–38.

LATE CRETACEOUS SEA-LEVEL CHANGE

Erickson, J. P., and J. L. Pindel. 1994. Sequence stratigraphy and relative sea-level history of the Cretaceous to Eocene passive margin of northeastern Venezuela, and the possible tectonic and eustatic causes of stratigraphic development. *Society of Economic Paleontologists and Mineralogists Special Publication*. In press.

Hallam, A. 1992. *Phanerozoic Sea-level Changes*. New York: Columbia University Press.

Hess, J., M. L. Bender, and J.-G. Schilling. 1986. Evolution of the ratio of strontium-87 to strontium-86 in seawater from Cretaceous to present. *Science* 231:979–84.

Officer, C. B. 1990. Extinctions, iridium, and shocked minerals associated with the Cretaceous/Tertiary transition. *Journal of Geological Education* 38:403–25.

K-T VOLCANISM

Browne, M. W. 1985. Dinosaur experts resist meteor extinction idea. *The New York Times*. 29 October:C1, C3–C4.

Browne, M. W. 1988. The debate over dinosaur extinctions takes an unusually rancorous turn. *The New York Times*. 19 January:C1, C4.

Browne, M. W. 1992. New clues to agent of life's worst extinction, *The New York Times*. 15 December:C1, C13.

Campbell, I. H., G. K. Czamanske, V. A. Federenko, R. I. Hill, and V. Stepanov. 1992. Synchronism of the Siberian Traps and the Permian-Triassic boundary. *Science* 258:1760–63.

Courtillot, V. E. 1990. What caused the mass extinctions: A volcanic eruption. *Scientific American* 263, no. 4:85–92.

Courtillot, V. E., J. Besse, D. Vandamme, R. Montigny, J.-J. Jaeger, and H. Capetta. 1986. Deccan flood basalts at the Cretaceous/Tertiary boundary. *Earth and Planetary Science Letters* 80:361–74.

Duncan, R. A., and D. G. Pyle. 1988. Rapid eruption of the Deccan flood basalts at the Cretaceous/Tertiary boundary. *Nature* 333:841–43.

Hallam, A. 1987. End-Cretaceous mass extinction event: Argument for terrestrial causation. *Science* 238:1237–42.

Keith, M. L. 1980. Cretaceous volcanism and the disappearance of the dinosaurs. *Eos* 61:400.

McLean, D. M. 1978. A terminal Mesozoic "greenhouse": Lessons from the past. *Science* 201:401–406.

Officer, C. B., A. Hallam, C. L. Drake, and J. D. Devine. 1987. Late Cretaceous and paroxysmal Cretaceous/Tertiary extinctions. *Nature* 326:143–49.

Tweto, O. 1975. Laramide (Late Cretaceous-Early Tertiary) orogeny in the Southern Rocky Mountains. *Geological Society of America Memoir*. 144:1–44.

ANATOMY OF THE EXTINCTION DEBATE

Thomson, K.S. 1988. Anatomy of the extinction debate. *American Scientist* 76:59–61.

SIX *Science and Politics*

OBLOQUY

Browne, M. W. 1988. The debate over dinosaur extinctions takes an unusually rancorous turn. *The New York Times.* 19 January:C1, C4.

Davis, N. P. 1968. *Lawrence and Oppenheimer.* New York: Simon and
Schuster.

NUCLEAR WINTER AND PUNCTUATED EQUILIBRIUM

Dyson, F. J. 1988. *Infinite in All Directions.* New York: Harper and Row.

Gould, S. J., and N. Eldredge. 1977. Punctuated equilibria: The tempo
and mode of evolution reconsidered. *Paleobiology* 3:115–51.

Turco, R. P., O. B. Toon, T. P. Ackerman, J. B. Pollack, and C. Sagan,
1983. Nuclear winter: Global consequences of multiple nuclear explosions. *Science* 222:1283–92.

SEVEN *Media Science*

POLYWATER DISPUTE

Franks, F. 1982. *Polywater.* Cambridge: MIT Press.

MEDIA SCIENCE

Browne, M. W. 1985. Dinosaur experts resist meteor extinction idea.
New York Times. 29 October:C1, C4.

Hoffman, A., and M. H. Nitecki. 1985. Reception of the asteroid hypothesis of terminal Cretaceous extinctions. *Geology* 13:884–87.

Keller, G. 1994. K-T boundary issues. *Science* 264:641.

Kerr, R. A. 1989. Asteroids, dinosaurs and the big splat. *Washington Post.*
7 May:B3.

McDonald, K. A. 1992. New data suggesting an asteroid impact inflame
debate over dinosaurs' demise. *Chronicle of Higher Education.* 29 Oc-
tober:A7–A9.

Rice, A. 1989. Snowbird II: A dissenting view. *Science* 243:875–76.

EIGHT *Iridium and Shocked Minerals*

IRIDIUM IN THE GEOLOGICAL RECORD

Chyi, L. L. 1982. The distribution of gold and platinum in bituminous
coal. *Economic Geology* 77:1592–97.

Crocket, J. H. 1981. Geochemistry of the platinum-group elements. *Canadian Institute of Mining and Metallurgy Special Volume* 23:49–64.

Crocket, J. H., J. D. MacDougall, and R. C. Harris. 1973. Gold, palladium and iridium in marine sediments. *Geochimica et Cosmochimica Acta* 37:2547–56.

Felitsyn, S. B., and P. A. Vaganov. 1988. Iridium in the ash of Kamchatkan volcanoes. *International Geology Review* 30:1288–91.

Gostin, V. A., R. R. Keays, and W. W. Wallace. 1989. Iridium anomaly from the Acraman impact ejecta horizon: Impacts can produce sedimentary iridium peaks. *Nature* 340:542–44.

Koeberl, C. 1989. Iridium enrichment in volcanic dust from blue ice fields, Antarctica, and possible relevance to the K/T boundary event. *Earth and Planetary Science Letters* 92:317–22.

Toutain, J.-P., and G. Meyer. 1989. Iridium-bearing sublimates at a hot-spot volcano (Piston de la Fournaise, Indian Ocean). *Geophysical Research Letters* 16:1391–94.

Tredoux, M., M. J. DeWit, R. J. Hart, R. A. Armstrong, N. M. Lindsay and J.P.F. Sellschop. 1989. Platinum group elements in a 3.5 Ga nickel-iron occurrence: Possible evidence of a deep mantle origin. *Journal of Geophysical Research* 94:795–813.

Zoller, W. H., J. R. Parrington, and J. M. Phelan Kotra. 1983. Iridium enrichment in airborne particles from Kilauea volcano: January 1983. *Science* 222:1118–21.

IRIDIUM AT K-T SECTIONS

Crocket, J. H., C. B. Officer, F. C. Wezel, and G. D. Johnson. 1988. Distribution of noble metals across the Cretaceous/Tertiary boundary at Gubbio, Italy: Iridium variation as a constraint on the duration and nature of Cretaceous/Tertiary boundary events. *Geology* 16:77–80.

Michel, H. V., F. Asaro, W. Alvarez, and L. W. Alvarez. 1985. Elemental profile of iridium and other elements near the Cretaceous/Tertiary boundary in hole 577B. *Initial Reports of the Deep Sea Drilling Project* 86:533–38.

Officer, C. B. 1982. Mixing, sedimentation rates and age dating for sediment cores. *Marine Geology* 46:261–78.

Officer, C. B. 1992. The relevance of iridium and microscopic dynamic deformation features toward understanding the Cretaceous/Tertiary transition. *Terra Nova* 4:394–404.

Rocchia, R., D. Boclet, Ph. Bonté, J. Devineau, C. Jéhanno, and M. Renard. 1987. Comparison des distributions de l'iridium observées à la limite Crétacé/Tertiaire dans divers sites Européens. *Mémoire de la Société Géologique de France* 150:93–103.

Rocchia, R., D. Boclet, Ph. Bonté, C. Jéhanno, Y. Chen, V. Courtillot, C. Mary, and F. Wezel. 1990. The Cretaceous-Tertiary boundary at Gubbio revisited: Vertical extent of the Ir anomaly. *Earth and Planetary Science Letters* 99:206–19.

Wright, A. A., U. Bleil, S. Monechi, H. V. Michel, N. J. Shackleton, B.R.T. Simoneit, and J. C. Zachos. 1985. Summary of Cretaceous/ Tertiary boundary studies, Deep Sea Drilling Project Site 577, Shatsky Rise. *Initial Reports of the Deep Sea Drilling Project* 86:799–804.

OTHER ELEMENTAL ENHANCEMENTS AT K-T SECTIONS

De Paolo, D. J., K. T. Kyte, B. D. Marshall, J. R. O'Neil, and J. Smit. 1983. Rb-Sr, Sm-Nd, K-Ca, O, and H isotopic study of Cretaceous-Tertiary boundary sediments, Caravaca, Spain: Evidence for an oceanic impact. *Earth and Planetary Science Letters* 64:356–73.

Finnegan, D. L., T. L. Miller, and W. H. Zoller. 1990. Iridium and other trace-metal enrichments from Hawaiian volcanoes. *Geological Society of America Special Paper* 247:111–16.

Luck, J.-M., and K. K. Turekian. 1983. Osmium-187/osmium-186 in manganese nodules and the Cretaceous-Tertiary boundary. *Science* 222:613–15.

Officer, C. B., and C. L. Drake. 1985. Terminal Cretaceous environmental events. *Science* 227:1161–67.

Tredoux, M., M. J. DeWit, R. J. Hart, N. M. Lindsay, B. Verhagen, and J.P.F. Sellschop. 1989. Chemostratigraphy across the Cretaceous-Tertiary boundary and a critical assessment of the iridium anomaly. *Journal of Geology* 97:585–605.

PDF'S IN THE GEOLOGICAL RECORD

Alexopoulos, J. S., R.A.F. Grieve, and P. B. Robertson. 1988. Microscopic lamellar deformation features in quartz: Discriminative characteristics of shock generated varieties. *Geology* 16:796–99.

Bunch, T. E. 1968. Some characteristics of selected minerals from craters. In B. M. French and N. M. Short, eds., *Shock Metamorphism of Natural Materials*. Baltimore: Mono Book.

Carter, N. L. 1968. Dynamic deformation of quartz. In B. M. French and N. M. Short, eds., *Shock Metamorphism of Natural Materials*. Baltimore: Mono Book.

Carter, N. L., C. B. Officer, C. A. Chesner, and W. I. Rose. 1986. Dynamic deformation of volcanic ejecta from the Toba caldera: Possible relevance to Cretaceous/Tertiary boundary phenomena. *Geology* 14:380–83.

Heard, H. C., and N. L. Carter. 1968. Experimentally induced "natural" intragranular flow in quartz and quartzite. *American Journal of Science* 266:1–42.

Lyons, J. B., C. B. Officer, P. E. Borella, and R. Lahodynsky. 1993. Planar lamellar substructures in quartz. *Earth and Planetary Science Letters* 119:431–40.

Officer, C. B., and N. L. Carter. 1991. A review of the structure, petrology, and dynamic deformation characteristics of some enigmatic terrestrial structures. *Earth Science Reviews* 30:1–49.

Robertson, P. B., M. R. Dence, and M. A. Vos. 1968. Deformation in rock forming minerals from Canadian craters. In B. M. French and N. M. Short, eds., *Shock Metamorphism of Natural Materials*. Baltimore: Mono Book.

PDF's AT K-T SECTIONS

Bohor, B. F., E. E. Foord, P. J. Modreski, and D. M. Triplehorn. 1984. Mineralogic evidence for an impact event at the Cretaceous-Tertiary boundary. *Science* 224:867–69.

Carter, N. L., C. B. Officer, and C. L. Drake. 1990. Dynamic deformation of quartz and feldspar: Clues to causes of some natural crises. *Tectonophysics* 171:373–91.

Huffman, A. R., J. H. Crocket, N. L. Carter, P. E. Borella, and C. B. Officer. 1990. Chemistry and mineralogy across the Cretaceous/Tertiary boundary at DSDP site 527, Walvis Ridge, South Atlantic Ocean. *Geological Society of America Special Paper* 247:319–34.

Izett, G. A. 1990. The Cretaceous/Tertiary boundary interval, Raton Basin, Colorado and New Mexico. *Geological Society of America Special Paper* 249:1–100.

Izett, G. A., W. A. Cobban, J. D. Obradovich, and M. J. Kunk. 1993. The Manson impact structure: $^{40}Ar/^{39}Ar$ age and its distal impact ejecta in the Pierre Shale in Southeastern South Dakota. *Science* 262:729–32.

Quezada Muñeton, J. M., L. E. Marin, V. L. Sharpton, G. Ryder, and B. C. Shuraytz. 1992. The Chicxulub impact structure. *Lunar and Planetary Science Conference Abstracts* 23:1121–22.

Stinnesbeck, W., J. M. Barbarin, G. Keller, J. G. Lopez-Oliva, D. A. Pivnik, J. B. Lyons, C. B. Officer, T. Adatte, G. Graup, R. Rocchia, and E. Robin. 1993. Deposition of channel deposits near the Cretaceous-Tertiary boundary in northeastern Mexico: Catastrophic or normal sedimentary deposits. *Geology* 21:979–800.

Tschudy, R. H., and B. D. Tschudy. 1986. Extinction and survival of plant life following the Cretaceous/Tertiary boundary event, Western Interior, North America. *Geology* 14:667–70.

NINE *The Silly Season*

PERIODIC COMET SHOWERS

Angier, N. 1985. Did comets kill the dinosaurs? *Time* 6 May:72–83.

Clemens, E. S. 1986. Of asteroids and dinosaurs: The role of the press in the shaping of scientific debate. *Social Studies of Science* 16:421–56.

Davis, M., P. Hut, and R. A. Muller. 1984. Extinction of species by periodic comet showers. *Nature* 308:715–17.

Hallam, A. 1984. The causes of mass extinctions. *Nature* 308:686–87.

Hallam, A. 1986. The Pliensbachian and Tithonian extinction events. *Nature* 319:765–68.

Hoffman, A. 1985. Patterns of family extinction depend on definition and geological timescale. *Nature* 319:765–68.

Kitchell, J. A., and D. Pena. 1984. Periodicity of extinctions in the geologic past: Deterministic versus stochastic explanations. *Science* 226:689–92.

Kyte, F. T., and J. T. Wasson. 1986. Accretion rate of extraterrestrial matter: Iridium deposited 33 to 67 million years ago. *Science* 232:1225–29.

Maddox, J. 1985. Periodic extinctions undermined. *Nature* 315:627.

Muller, R. A. 1985. An adventure in science. *The New York Times Sunday Magazine* 24 March:33–44, 50.

Rampino, M. R., and R. B. Stothers. 1984. Terrestrial mass extinctions, cometary impacts and the Sun's motion perpendicular to the galactic plane. *Nature* 308:709–12.

Raup, D. M., and J. J. Sepkoski. 1984. Periodicity of extinctions in the geologic past. *Proceedings of the National Academy of Sciences* 81:801–805.

Stigler, S. M., and M. J. Wagner. 1987. A substantial bias in nonparametric tests for periodicity in geophysical data. *Science* 238:940–45.

Whitmire, D. P., and J. J. Matese. 1985. Periodic comet showers and planet X. *Nature* 313:36–38.

SEVERAL IMPACTS AT K-T TIME

Alvarez, W., and F. Asaro. 1990. What caused the mass extinction: An extraterrestrial impact. *Scientific American* 263, no. 4:78–84.

Hut, P., W. Alvarez, W. P. Elder, T. Hansen, E. G. Kaufman, G. Keller, E. M. Shoemaker, and P. R. Weisman. 1987. Comet showers as a cause of mass extinctions. *Nature* 329:118–26.

Officer, C. B. 1993. Death of the dinosaurs. *New Scientist* 20 February: 34–38.

GLOBAL WILDFIRE

Hansen, H. J., K. L. Rasmussen, R. Gwozdz, and H. Kuzendorf. 1987. Iridium-bearing carbon black at the Cretaceous-Tertiary boundary. *Geological Society of Denmark Bulletin* 36:305–314.

Joyce, C. 1985. Worldwide fire killed the dinosaurs, *New Scientist*, 10 October:26.

Officer, C. B., and A. A. Ekdale. 1986. Cretaceous extinctions and wildfires. *Science* 234:262–63.

Weisburd, S. 1985. Wildfires: Apocalypse then and now. *Science News* 12 October:228.

Wolbach, W. S., R. S. Lewis, and E. Anders. 1985. Cretaceous extinctions: Evidence for wildfires and search for meteoritic material. *Science* 230:167–170.

Wolbach, W. S., I. Gilmour, E. Anders, C. J. Orth, and R. R. Brooks. 1988. Global wildfire at Cretaceous-Tertiary boundary. *Nature* 334:665–69.

AMINO ACIDS

Cronin, J. R. 1989. Amino acids and bolide impacts. *Nature* 339:423–24.

Olson, E. S. 1992. Amino acids from coal gasification. *Nature* 357:202.

Zahnle, K., and D. Grinspoon. 1990. Comet dust as a source of amino acids at the Cretaceous/Tertiary boundary. *Nature* 348:157–60.

Zhao, M., and J. L. Bada. 1989. Extraterrestrial amino acids in Cretaceous/Tertiary boundary sediments at Stevns Klint, Denmark. *Nature* 339:463–65.

IMPACTS AND VOLCANISM

Courtillot, V. 1990. What caused the mass extinctions: A volcanic eruption. *Scientific American* 263, no. 4:85–92.

Loper, D. E., and K. McCartney. 1990. On impacts as a cause of geomagnetic field reversals or flood basalts. *Geological Society of America Special Paper* 247:19–25.

Rampino, M. R., and R. B. Stothers. 1988. Flood basalt volcanism during the past 250 million years. *Science* 241:663–68.

TEN *The Missing Crater*

CUBA GEOLOGIC SECTION

Bohor, B. F., and R. Seitz. 1990. Cuban K/T catastrophe. *Nature* 344:593.

Dietz, R. S., and J. McHone. 1990. Isle of Pines (Cuba), apparently not K/T boundary impact site. *Geological Society of America Abstracts* 22, no. 7:A79.

Iturralde-Vinent, M. A. 1992. A short note on the Cuban late Maastrichtian megaturbidite (an impact-derived deposit?). *Earth and Planetary Science Letters* 109:225–28.

HAITI GEOLOGIC SECTION

Blum, J. D., and C. P. Chamberlain. 1992. Oxygen isotope constraints on the origin of impact glasses from the Cretaceous-Tertiary boundary. *Science* 257:1104–1107.

Hildebrand, A. R., and W. V. Boynton. 1990. Proximal Cretaceous-Tertiary boundary impact deposits in the Caribbean. *Science* 248:843–47.

Izett, G. A. 1991. Tektites in Cretaceous-Tertiary boundary rocks on Haiti and their bearing on the Alvarez impact hypothesis. *Journal of Geophysical Research* 96:20,879–905.

Jéhanno, C., D. Boclet, L. Froget, B. Lambert, E. Robin, R. Rocchia, and L. Turpin. 1992. The Cretaceous-Tertiary boundary at Beloc, Haiti: No evidence for an impact in the Caribbean area. *Earth and Planetary Science Letters* 109:229–41.

Lyons, J. B., and C. B. Officer. 1992. Mineralogy and petrology of the Haiti Cretaceous/Tertiary section. *Earth and Planetary Science Letters* 109:205–24.

Maurrasse, F.J.-M.R., F. Pierre-Louis and J.J.-G. Rigaud. 1985. Upper Cretaceous to lower Paleocene pelagic calcareous deposits in the southern peninsula of Haiti: Their bearing on the problem of the Cretaceous/Tertiary boundary. *Latin American Geological Congress Transactions* 4:328–37.

Officer, C. B., and J. B. Lyons. 1993. A short note on the origin of the yellow glasses at the Haiti Cretaceous/Tertiary section. *Earth and Planetary Science Letters* 118:349–51.

Sigurdsson, H., S. D'Hondt, M. A. Arthur, T. J. Bralower, J. C. Zachos, M. van Fossen, and J.E.T. Channell. 1991. Glass from the Cretaceous/Tertiary boundary in Haiti. *Nature* 349:482–87.

DSDP SITE 536 AND 540 GEOLOGIC SECTIONS

Alvarez, W., J. Smit, W. Lowrie, F. Asaro, S. V. Margolis, P. Claeys, M. Kastner, and A. R. Hildebrand. 1992. Proximal impact deposits at the Cretaceous-Tertiary boundary in the Gulf of Mexico: A re-study of DSDP Leg 77 Sites 536 and 540. *Geology* 20:697–770.

Buffler, R. T., W. Schlanger et al. 1984. Sites 535, 539, and 540. *Initial Reports of the Deep Sea Drilling Project* 77:25–217.

Buffler, R. T., W. Schlanger et al. 1984. Site 536. *Initial Reports of the Deep Sea Drilling Project* 77:219–54.

Keller, G., N. MacLeod, J. B. Lyons, and C. B. Officer. 1993. Is there evidence for Cretaceous-Tertiary boundary-age deep-water deposits in the Caribbean and Gulf of Mexico? *Geology* 21:776–80.

MIMBRAL, MEXICO, GEOLOGIC SECTION

Muir, J. M. 1936. *Geology of the Tampico Region.* Tulsa: American Association of Petroleum Geologists.

Smit, J., A. Montanari, N.H.M. Swinburne, W. Alvarez, A. R. Hildebrand, S. V. Margolis, P. Claeys, W. Lowrie, and F. Asaro. 1992. Tektite-bearing, deep-water clastic unit at the Cretaceous-Tertiary boundary in northeastern Mexico. *Geology* 20:99–103.

Stinnesbeck, W., J. M. Barbarin, G. Keller, J. G. Lopez-Oliva, D. A. Pivnik, J. B. Lyons, C. B. Officer, T. Adatte, G. Graup, R. Rocchia, and E. Robin. 1993. Deposition of channel deposits near the Cretaceous-Tertiary boundary in northeastern Mexico: Catastrophic or "normal" sedimentary deposits. *Geology* 21:797–800.

HAITI, CUBA, AND COLOMBIAN BASIN STRUCTURES

Bohor, B. F., and R. Seitz. 1990. Cuban K/T catastrophe. *Nature* 344:593.

Dietz, R. S., and J. McHone. 1990. Isle of Pines (Cuba), apparently not K/T boundary impact site. *Geological Society of America Abstracts* 22, no. 7:A79.

Hildebrand, A. R., and W. V. Boynton. 1990. Proximal Cretaceous-Tertiary boundary impact deposits in the Caribbean. *Science* 248:843–47.

Maurrasse, F.J-M.R. 1990. The Cretaceous-Tertiary boundary impact site in the Caribbean. *Geological Society of America Abstracts* 22, no. 7:A77.

CHICXULUB, MEXICO, STRUCTURE

Anonymous. 1994. The puzzle of Chicxulub and Chicxulub momentum. *Geology Today* 10:122.

Hildebrand, A. R., and W. V. Boynton. 1990. Proximal Cretaceous-Tertiary boundary impact deposits in the Caribbean. *Science* 248:843–47.

Hildebrand, A. R., G. T. Penfield, D. A. Kring, M. Pilkington, A. Camargo Z., S. G. Jacobsen, and W. V. Boynton. 1991. Chicxulub crater: A possible Cretaceous/Tertiary boundary impact crater on the Yucatán Peninsula, Mexico. *Geology* 19:867–71.

Lopez-Ramos, E. 1975. Geological summary of the Yucatán peninsula. In A.E.M. Nairn and F. G. Stehli, eds., *The Ocean Basins and Margins*, vol. 3, *The Gulf of Mexico and the Caribbean*. New York: Plenum Press.

Melosh, H. J. 1989. *Impact Cratering*. New York: Oxford University Press.

Meyerhoff, A. A., J. B. Lyons, and C. B. Officer. 1994. Chicxulub structure: A volcanic sequence of Late Cretaceous age. *Geology* 22:3–4.

Officer, C. B. 1994. Chicxulub structure: A volcanic sequence of Late Cretaceous age. *Paleontological Society of America Special Publication* 7:425–36.

Officer, C. B., C. L. Drake, J. L. Pindell, and A. A. Meyerhoff. 1992. Cretaceous-Tertiary events and the Caribbean caper. *GSA Today* 2:69–75.

Penfield, G. T., and A. Camargo. 1981. Definition of a major igneous zone in the central Yucatan platform with aeromagnetics and gravity. *Society of Exploration Geophysicists Annual Meeting Abstracts* 51:37.

Quezada Muñeton, J. M., L. E. Marin, V. L. Sharpton, G. Ryder, and B. C. Schuraytz. 1992. The Chicxulub impact structure: Shock deformation and target composition. *Lunar and Planetary Science Conference Abstracts* 23:1121–22.

Sharpton, V. L., G. B. Dalrymple, L. E. Marin, G. Ryder, B. C. Schuraytz, and J. Urrutia-Fucugauchi. 1992. New links between the Chicxulub impact structure and the Cretaceous/Tertiary boundary. *Nature* 359:819–21.

Swisher, C. C., J. M. Grajales-Nishimura, A. Montanari, S. V. Margolis, P. Claeys, W. Alvarez, P. Renne, E. Cedillo-Pardo, F.J.-M.R. Maurrasse, G. H. Curtis, J. Smit, and M. O. McWilliams. 1992. Coeval ^{40}Ar/^{39}Ar ages of 65.0 million years ago from Chicxulub crater melt rock and Cretaceous-Tertiary boundary tektites. *Science* 257:954–58.

ELEVEN *What Did Happen*

PLATE TECTONICS

Anderson, D. L. 1962. The plastic layer of the earth's mantle. *Scientific American* 207, no. 1:52–59.

Officer, C. B., and C. L. Drake. 1983. Plate dynamics and isostasy in a dynamic system. *Journal of Geophysics* 54:1–19.

Uyeda, S. 1971. *The New View of the Earth*. San Francisco: W.H. Freeman.

MANTLE PLUMES

Kennett, J. P. 1981. Marine tephrochronology. In C. Emiliani, ed., *The Sea*, vol. 7. New York: Wiley-Interscience.

Loper, D. E., K. McCartney, and G. Buzyna. 1988. A model of correlated episodicity in magnetic-field reversals, climate, and mass extinctions. *Journal of Geology* 96:1–15.

Vink, G. E., W. J. Morgan, and P. R. Vogt. 1985. The earth's hot spots. *Scientific American* 252, no. 4:50–57.

Vogt, P. R. 1972. Evidence for global synchronism in mantle plume convection, and possible significance for geology. *Nature* 240:338–42.

DECCAN AND OTHER VOLCANISM AT K-T TIME

Borella, P. E. 1984. Sedimentology, petrology and cyclic sedimentation patterns, Walvis Ridge transect, Leg 74, Deep Sea Drilling Project. *Initial Reports of the Deep Sea Drilling Project* 74:645–62.

Courtillot, V. E. 1990. What caused the mass extinctions: A volcanic eruption. *Scientific American* 263, no. 4:85–92.

Tweto, O. 1975. Laramide (Late Cretaceous-Early Tertiary) orogeny in the Southern Rocky Mountains. *Geological Society of America Memoir* 144:1–44.

P-TR AND K-T COMPARISONS

Browne, M. W. 1992. New clues to agent of life's worst extinction. *The New York Times*, 15 December:C1, C3.

Campbell, I. H., G. K. Czamanske, V. A. Federenko, R. I. Hill, and V. Stepanov. 1992. Synchronism of the Siberian Traps and the Permian-Triassic boundary. *Science* 258:1760–63.

Holser, W. T., and M. Magaritz. 1987. Events near the Permian-Triassic boundary. *Modern Geology* 11:155–80.

Holser, W. T., H.-P. Schönlaub, M. Attrep, Jr., K. Boeckelmann, P. Klein, M. Magaritz, C. J. Orth, A. Fenninger, C. Jenny, M. Kralik, H. Mauritsch, E. Pak, J.-M. Schramm, K. Statteger, and

R. Schmöller. 1989. A unique geochemical record at the Permian/ Triassic boundary. *Nature* 337:39–44.

K-T ENVIRONMENTAL CHANGES

Archibald, J. D., and W. A. Clemens. 1982. Late Cretaceous extinctions. *American Scientist* 70:377–85.

Hallam, A. 1987. End Cretaceous mass extinction event: Argument for terrestrial causation. *Science* 238:1237–42.

Officer, C. B., A. Hallam, C. L. Drake, and J. D. Devine. 1987. Late Cretaceous and paroxysmal Cretaceous/Tertiary extinctions. *Nature* 326:143–49.

SELENIUM

Anderson, M. S., H. W. Larkin, K. C. Beeson, F. F. Smith, and E. Thacker. 1961. Selenium in agriculture. *United States Department of Agriculture Handbook* 200.

Koeberl, C. 1989. Iridium enrichment from blue ice fields, Antarctica, and possible relevance to the K/T boundary event. *Earth and Planetary Science Letters* 92:317–22.

Rosenfeld, I., and O. A. Beath. 1964. *Selenium: Geobotany, Biochemistry, Toxicity, and Nutrition.* New York: Academic Press.

Smit, J., and W.G.H.Z. ten Kate. 1992. Trace-element patterns at the Cretaceous-Tertiary boundary—consequences of a large impact. *Cretaceous Research* 3:307–22.

Zoller, W. H., J. R. Parrington, and J. M. Phelan Kotra. 1983. Iridium enrichment in airborne particles from Kilauea volcano: January, 1983. *Science* 222:1118–21.

AFTERWORD

PROGRESSIVE AND DEGENERATIVE RESEARCH PROGRAMMES

Alvarez, W., and F. Asaro. 1990. What caused the mass extinction: An extraterrestrial impact. *Scientific American* 263, no. 4:78–84.

Carter, N. L., C. B. Officer, C. A. Chesner, and W. I. Rose. 1986. Dynamic deformation of volcanic ejecta from the Toba caldera: Possible relevance to Cretaceous/Tertiary boundary phenomena. *Geology* 14:380–83.

Hut, P., W. Alvarez, W. P. Elder, T. Hansen, E. G. Kaufmann, G. Keller, E. M. Shoemaker, and P. R. Weisman. 1987. Comet showers as a cause of mass extinctions. *Nature* 329:118–26.

Lakatos, I. 1978. *The Methodology of Scientific Research Programmes*. Cambridge: Cambridge University Press.

Zoller, W. H., J. R. Parrington, and J. M. Phelan Kotra. 1983. Iridium enrichment in airborne particles from Kilauea volcano: January, 1983. *Science* 222:1118–21.

PATHOLOGICAL SCIENCE

Langmuir, I. (transcribed and edited by R. N. Hall). 1989. Pathological science. *Physics Today* 42, no. 10:36–48.

Officer, C. B. 1993. Death of the dinosaurs. *New Scientist* 20 February: 34–38.

Rousseau, D. L. 1992. Case studies in pathological science. *American Scientist* 80, no. 1:54–63.

INDEX